Fire Assaying Gold, Silver and Lead Ores

by L.S. Austin

with an introduction by Kerby Jackson

This work contains material that was originally published in 1907 by the US Department of Interior.

This publication was created and published for the public benefit, utilizing public funding and is within the Public Domain.

This edition is reprinted for educational purposes and in accordance with all applicable Federal Laws.

Introduction Copyright 2014 by Kerby Jackson

Introduction

It has been over a hundred years since the Department of Interior released it's important publication "The Fire Assay of Gold, Silver and Lead in Ores and Metallurgical Products". First released in 1907, this important volume has now been out of print for over a century and has been unavailable to the mining community since those days, with the exception of expensive original collector's copies and poorly produced digital editions.

It has often been said that "*gold is where you find it*", but even beginning prospectors understand that their chances for finding something of value in the earth or in the streams of the Golden West are dramatically increased by going back to those places where gold and other minerals were once mined by our forerunners. Despite this, much of the contemporary information on local mining history that is currently available is mostly a result of mere local folklore and persistent rumors of major strikes, the details and facts of which, have long been distorted. Long gone are the old timers and with them, the days of first hand knowledge of the mines of the area and how they operated. Also long gone are most of their notes, their assay reports, their mine maps and personal scrapbooks, along with most of the surveys and reports that were performed for them by private and government geologists. Even published books such as this one are often retired to the local landfill or backyard burn pile by the descendents of those old timers and disappear at an alarming rate. Despite the fact that we live in the so-called "Information Age" where information is supposedly only the push of a button on a keyboard away, true insight into mining properties remains illusive and hard to come by, even to those of us who seek out this sort of information as if our lives depend upon it. Without this type of information readily available to the average independent miner, there is little hope that our metal mining industry will ever recover.

This important volume and others like it, are being presented in their entirety again, in the hope that the average prospector will no longer stumble through the overgrown hills and the tailing strewn creeks without being well informed enough to have a chance to succeed at his ventures.

Kerby Jackson
Josephine County, Oregon
July 2014

TABLE OF CONTENTS

		PAGE
	Preface	7
I.	The Fire Assay of Ores or Furnace and Mill Products Containing Gold, Silver, and Lead	11
II.	Sampling and Preparation of the Ore for Assay	12
III.	Care of the Assay Office	15
IV.	Apparatus	17
V.	The Assay Furnace	22
VI.	Crucibles and Scorifiers	30
VII.	Assay Balances	32
VIII.	Fluxes Used in Assaying	40
IX.	Ores	48
X.	The Scorification Assay	50
XI.	Cupelling	54
XII.	Parting	56
XIII.	The Crucible Assay	60
XIV.	Roasting of Ores	70
XV.	Assay of Matte	72
XVI.	High-Grade Silver-Sulphide Assay	73
XVII.	Assay of Cyanide Solutions	74
XVIII.	Assay of Base-Bullion	76
XIX.	Assay of Silver Bars or Ingots	78
XX.	Assay of Blister or Pig-Copper Containing Silver and Gold	82
XXI.	Assay of Gold Bullion	84
XXII.	Assay of Ores Containing Metallics	85
XXIII.	The Lead Assay	87

PREFACE

The following pages present a system of assaying, intended to cover determinations of the precious metals (silver and gold) and of lead, according to methods quite commonly recognized in the Rocky Mountain States as suited to ores and products containing the just-named metals.

In reading books on the subject, the beginner in assaying is embarrassed in deciding which one of the various prescribed charges he is to employ. It has, therefore, been the aim of the author in this book to carry through a single system, which, while not the only one, would at least be definite and clear to the learner.

I.—THE FIRE ASSAY OF ORES OR FURNACE AND MILL PRODUCTS CONTAINING GOLD, SILVER, AND LEAD.

The fire assay is a dry or fusion method for the determination of the metal sought with the aid of heat and suitable fluxes, and of weighing it in metallic form. Thus if we desire to determine lead in an ore we fuse a known weight of the ore in a crucible together with suitable fluxes and obtain a lead button which, upon weighing, will give us the percentage of contained lead in the portion of ore so taken, called the sample. This sample must be a true average of the lot of ore whose value we wish thus to determine.

The steps in assaying are:
1. Receiving and labelling the ore.
2. Sampling.
3. Fluxing, melting, and cupelling.
4. Weighing the metal obtained.
5. Recording and reporting results.

Generally but one metal is determined in an assay, except that in the case of gold and silver, the two metals, recovered as one, are subsequently parted.

The work should be systematically arranged and carried out. The secret of doing a day's work well, both in quantity and in accuracy, depends largely upon this.

Never report results where there is reason to believe that they may be incorrect owing to faulty manipulation, to losses, or to the mixing of samples.

Single assays may be made upon grab samples when an approximation is sufficient, but for results involving the purchase of metals or ores, duplicates should be run which must closely agree, and, if not, then the work must be repeated until they do so.

II.—SAMPLING AND PREPARATION OF THE ORE FOR ASSAY.

All metals, ores, furnace and mill-products, which are to be assayed, must first be accurately sampled. This is quite as important as accurate assaying, for, if the sample does not truly represent the lot or quantity of ore whose value is sought, the assay of it will be misleading.

The following instructions on sampling ores are here given, since the assayer is often called upon to sample the ore which he is about to assay.

RECEIVING THE SAMPLE. The ore, mineral, or metallurgical product may come as a hand-sample, a 'grab'-sample, or as a regular lot-sample from the sampling mill. Only regular or mill-samples are, in general, assayed in duplicate.

Small tin or graniteware pans are used to receive hand-samples, each being placed in a separate pan, and with a label, upon which is marked the name of the owner, the name of the ore, and the lot-number of the sample. Where the assayer has to number the lots himself, he should also describe the ore so that it may be distinguished again by the sender.

It is a good practice to reserve one or more of the larger pieces, after having partly broken down a sample, since in this way we may be better able to designate or otherwise recognize the ore.

The hand-sample may consist of one or of several lumps, of a mixture of fine and of lumpy ore, or of a regularly ground sample of ore ready for assaying. In the latter case, the assayer should be especially on his guard to exercise due care, since another portion of the same 'pulp' can be used as a check on his work at another time or by another assayer.

PREPARATION OF THE SAMPLE. A general principle in sampling is that the ratio between the larger pieces in the sample and the weight of the sample should not exceed a certain amount.

This is attained by breaking the pieces finer as the bulk of the sample is decreased.

The breaking up of the lumps of a sample may be accomplished by means of a laboratory crusher, by pestle and mortar, or by breaking on the grinding-plate with a hammer or muller. In the latter case it is well to provide a wooden frame, four inches high, and but two inches high at the front, which encloses the area of the plate, and which prevents the ore from escaping while being broken up. When the quantity of the sample is considerable, some time and labor may be saved by not breaking so finely at first, but by breaking smaller each time the sample is cut down.

The work of cutting down or of quartering may be performed as follows: The ore is made up into a coned pile in the centre of the plate, the material being poured or thrown upon the apex of the pile so as to effect an even distribution of fine and coarse. The cone is now flattened out to a circular cake or disk by scraping the ore outward every way from its apex. Two lines are marked at right angles, dividing the mass into quarters. The two opposite quarters are then removed, rejected, and the remainder mixed and quartered as before, taking care, however, first to break the pieces smaller if necessary.

Finally, as this operation is continued, there remain two to four ounces of the ore. This is ground to pass an 80-mesh screen. Where the ore is rich, spotty, or contains coarse particles of gold, it is better to use a 120-mesh screen. The sample must now be dried, if not already so. The final mixing of the sample, after it has been ground up, is done by 'rolling' it on a piece of oil-cloth or of rubber cloth. This is performed by laying or folding over the cloth so as to cause the ore to travel from side to side. It is then rolled at right angles to its former direction, and thus mixing proceeds until the operator is sure that the ore is quite uniform in all its parts. A beginner is very apt not to move or roll the ore sufficiently to insure its complete turning over. For practice, it would be well for him to mix some argols and soda in this way, until the absolutely uniform color of the mixture indicates that the mixing is complete.

An expeditious and satisfactory way of cutting down a sample consists in the use of the tin sampling-trough or riffle. The ore is scattered into the troughs from a shovel in such wise as not to overtop the troughs. Half of the ore remains in the riffle, and half drops through. When the troughs are full the ore in them is reserved, and more is again added, and so on, until the sample has been all put through. This reserved portion again goes through the riffle, thus again halving it. Of course when finer breaking is needed it must be done.

Prospectors and others, interested in mines, often deceive themselves by thinking that by sending in a piece or lump of rock, they can arrive at an idea of the value of the ore they are going to ship. If one will assay separately two such pieces of rock, he will see how much they vary from one another. The only way to know the value of an ore is to sample it in quantity, so that all parts of the pile are equally represented.

III.—CARE OF THE ASSAY OFFICE.

When the work of sweeping out or cleaning the assay office is done by an attendant, the assayer should take notice that it is thoroughly done. The balances, however, to keep them clean and in the best of order, need his personal attention. As the assayer's work proceeds through the day, he will find that dust, ashes, and scattering particles get on the desk and about the furnace. He has to keep his work going steadily, but there are spare moments when he can brush up and set things in order again. In fact, he should put his tools and appliances in place again as he goes along, and thus, nothing being mislaid, valuable time is saved.

It assists in the prestige of the assayer to keep his office constantly in order. If this is done, his employer, whose good-will he values, will notice these things in a way favorable to him; indeed it is one of the ways of showing that he is ambitious and energetic.

Clinkers will form upon the interior of the furnace, and these are to be removed with a long-handled cutter-bar or chisel. The assayer also replaces his own muffles, claying up the joint with fire-clay, to which has been added its own weight of sand, or preferably, some coarsely ground fire-brick which he can break up for himself on the grinding-plate. In a reduction works where tailing, slag, or low-grade products are treated, there should be two cast-iron bucking plates, one of which should be reserved for this use. In case a high-grade ore is followed by a low-grade one the assayer must use some sand, coke, or other similar material to be ground on the plate and then brushed off, in order to insure the entire removal of particles of the rich ore. Otherwise, low-grade samples may get 'salted.' This may account for cases where a trace, or even more of metal, has been reported when in fact the material was quite barren.

When auriferous placer gravel is to be assayed, a considerable sample of a known weight, say 100 assay tons (6.38 lb.), should be panned down carefully and the residual concentrate, consisting of black sand, pyrites, and gold, assayed. Even where the gravel has been broken up and mixed, the results are apt to be uncertain, since particles of gold are often lodged in the crevices of the pebbles, and only water treatment will wash them out.

IV.—APPARATUS.

GRINDING PLATE AND MULLER. For breaking, quartering down, and pulverizing samples of ore the grinding or bucking plate is employed, as shown in Fig. 1. It is of cast-iron, planed on its upper surface, one inch thick, and has raised edges. The waste part of the sample is swept into a box placed beneath the bench. The muller, which should weigh 25 lb., has a curved rubbing surface, and is rubbed back and forth on the plate to pulverize the ore. Where the pieces are one-fourth inch or larger, they are broken, using the muller as a hammer. Quartering down

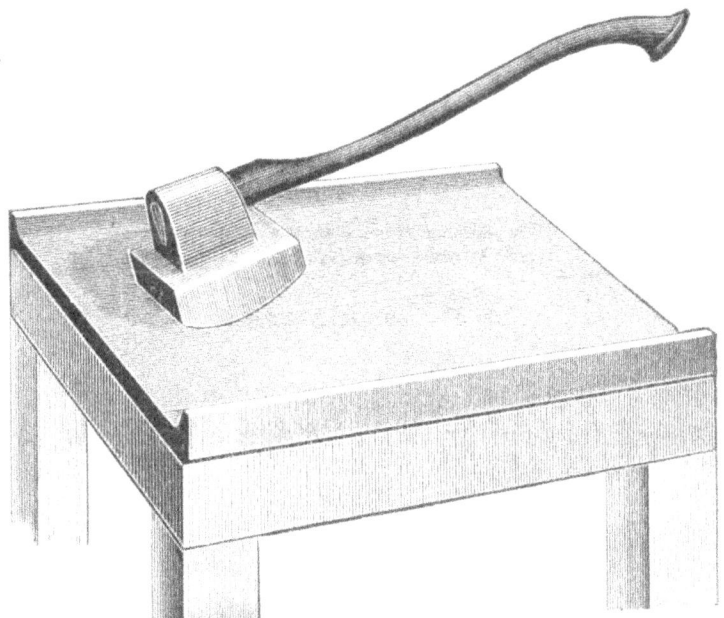

FIG. 1. GRINDING PLATE AND MULLER.

of the sample is performed on the plate using a piece of Russia sheet iron 3 by 6 in. as a scraper.

Sieves or screens may be 10 in. diam. with brass screen-cloth and having a removable pan-bottom (See Fig. 2). When the sample has been pulverized it is mixed or rolled on a piece of oil-cloth or a rubber cloth 18 in. square. Neglect of this precaution of thorough mixing means varying results in duplicate assays.

FIG. 2. TEN-INCH SIEVE AND BOTTOM.

FIRE TOOLS. For handling crucibles, etc., several kinds of tools are used as shown below:

Crucible tongs are employed for setting, removing, and pour-

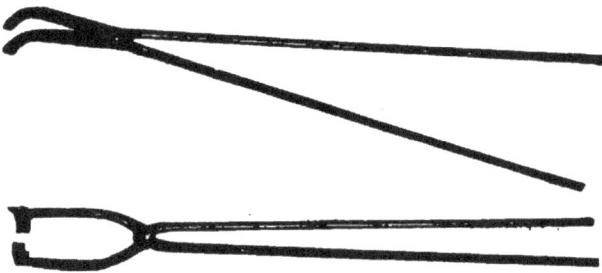

FIG. 3. CRUCIBLE TONGS.

ing crucibles. They have bent points for convenience in seizing the crucible and for pouring them. A pair of such tongs is shown in Fig. 3.

FIG. 4. SHORT CRUCIBLE TONGS.

Fig. 4 shows the short tongs used for removing nails from the crucible before pouring. They are nine inches long.

CRUCIBLE TONGS FOR MUFFLE. (See Fig. 5.) These have curved ends for embracing the crucible and are useful in putting in and removing crucibles (sometimes scorifiers) from the muffle. They have in mid-length a flat tongue for steadying the two legs in relation to one another. The round wire tongue, ordinarily provided, is not satisfactory.

FIG. 5. CRUCIBLE TONGS FOR MUFFLE USE.

For handling scorifiers we have scorifier-tongs, of which Fig. 6 is a good example. It has a forked leg as well as a flat one, and has a broad spring at the back. The flat end is often used to dip up and transfer borax-glass to the scorifier.

FIG. 6. SCORIFIER TONGS.

Fig. 7 represents a pair of cupel-tongs more carefully made and proportioned than those commonly purchased from dealers in assay supplies. It has a broad and flexible spring back, and a

FIG. 7. CUPEL TONGS.

flat tongue for steadying the legs. The bent points should be no longer than here given. The assayer can easily cut off and shape the points when, as is usual, they are too long. For large muffles they should be at least 30 in. long.

MUFFLE SCRAPER. This tool, intended for promptly scraping out anything spilled upon the floor of the muffle, should hang near the muffle so that it can be used without a moment's delay. In this way the damage to the muffle can be greatly lessened. It is shown in Fig. 8.

FIG. 8. SCRAPER.

CUPEL MOLDS. Cupels are made by the assayer (though they can also be bought) from bone-ash, ground fine enough to give a smooth surface and yet have a porous texture. The molds, 1½ in. diam., are in two parts, the ring and the plunger, and are generally made of brass. The assayer must take care to handle them so as not to bruise or deface them in any way. It will be noticed that the ring has an interior taper for the easy removal

FIG. 9. TWO-HOLE SLAG MOLD.

of the cupel when molded, and, in making the cupel, the plunger enters the smaller diameter. In making cupels, the bone-ash is

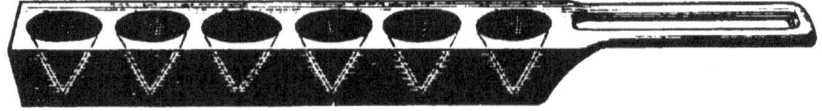

FIG. 10. SIX-HOLE SLAG MOLD.

moistened with a little water until it has a grip or adhesion such that, when molded, it will cohere. Before use, cupels must be carefully dried by being put in a warm place, preferably for a month or more, so that quite a stock must be made and kept on hand. There are cupel-making machines which will turn out several hundred cupels in an hour, but for a small assay office, they are generally made in the simpler way.

SLAG MOLDS. They are made of various shapes, some having a hemispherical, some a conical, cavity. The most satisfactory mold is one having the conical cavity, smoothly turned out, and in a solid cast-iron block, so as quickly to abstract the heat from the molten material poured into it. (See Fig. 9.) Fig. 10 represents a similar mold having six depressions.

Lighter and cheaper molds, having less iron in them, are not so satisfactory.

V.—THE ASSAY FURNACE.

A variety of assay furnaces have been devised, some using soft coal as a fuel and depending upon the flame from the coal to do the heating, some, in which coke is used, with coals in direct contact with the muffle or the crucible, and some where the flame of gasoline is used, much on the same principle as with soft coal. To these may be added ordinary illuminating gas, where, as in cities, it may be had.

Fig. 11. Coal-Burning Assay Furnace.

Fig. 11 is an elevation, and Fig. 12 and 13, two sectional elevations of a two-high muffle furnace for burning soft coal, which must have enough volatile constituents to produce a suitable flame. The interior is lined with fire-clay tile of special shapes so that the furnace can be readily and speedily built. The rest of the furnace is of red brick, the whole being firmly clamped or bound with angle-iron and tie rods. The muffles, of the form shown in Fig. 12 and 13, have an inside width of 9 in., and each

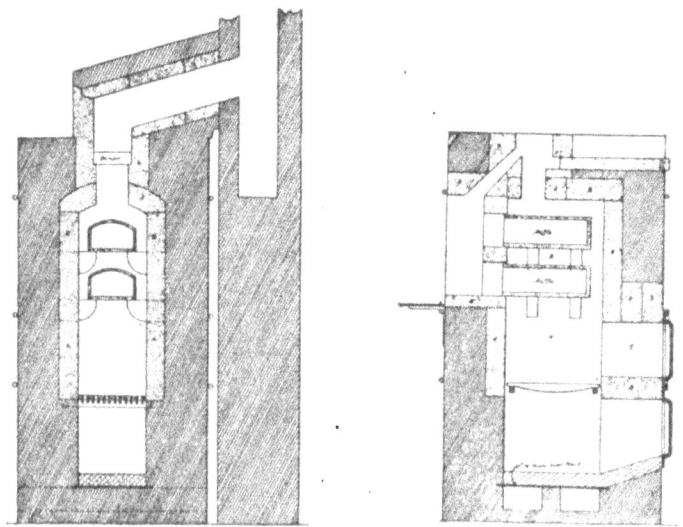

FIG. 12 AND 13. SECTIONS OF COAL-BURNING ASSAY FURNACE.

one is large enough to take two 20-gram crucibles (3 in. diam. by 3¾ in. high) abreast, and in at least four rows, making twelve crucibles at a time or five rows of scorifiers, four abreast, or easily four cupels abreast. The lower muffle, which can be brought to the higher temperature, is generally reserved for melting and the upper one for cupelling. By means of a tile-damper the assayer can regulate the heat of the furnace, so that, in cupelling, he can carry several rows of cupels at a time. He then subdues the heat at the back of the muffle by putting in cold crucibles which absorb the heat and cool down the muffle to exactly the required temperature. The muffles are carried on supports, two on either side. At the front will be noticed a short flue for removing the fumes which may issue from the muffles. The furnace should have a stack with a strong draft, this stack being built independently of the furnace. The fire-door is placed so as to feed the coal at the back, and out of the way of the assayer. It can also be arranged so as to feed at the side if so desired.

With a slight change in the fire-box, this furnace can burn wood, there being then but a single muffle used. These furnaces take 16 to 24-in. wood as desired.

Where much work has to be done, furnaces having much larger muffles are used, where as many as four 20-gm. crucibles can be put in abreast, five rows being melted at a time.

In firing with coal the bed should be not more than 6 in. deep, a little deeper at the back, using but little coal at a time and feed-

Fig. 14. Brown Portable Assay Furnace.

ing frequently so as to keep up a steady flame. Be careful not to permit any holes through the fuel bed by which cold air can get up to cool the muffle. The fire must be kept clean and free from ashes and cinders. As those latter accumulate in the ash-pit they must be removed. Otherwise, the heat from them softens and melts the grate bars, thus soon destroying them.

Fig. 14 is a perspective view of the well-known Brown portable furnace for coke. The pipe coming from it is 5 in. diam., and can be carried through the roof of an assay office, so that one can be quickly set up and put in operation. It contains a single muffle.

The muffle is suited to scorifying and cupelling, but not to receiving crucibles. Crucibles are, accordingly, set among the coke through the top door or lid. The crucibles are set in two rows parallel to the muffle, three pointed E crucibles in a row. They are set to leave 1.5 in. between them and the side of the furnace, and are packed all around with coke to their tops. It is upon the thorough performance of this work that the success of the melting depends. The fire is lit with a little paper and some light kindling followed by heavier pieces at least an inch thick and packed under and up to the muffle. Charcoal is then put in to

FIG. 15. BOSWORTH ASSAY FURNACE.

the level of the top of the muffle and then two inches in depth of coke broken to egg size or less. If left larger, these pieces will not pass down between the muffle and the wall of the furnace. The fire is then lit, and, when the wood has burned away, is poked down with a poker $\frac{3}{8}$ in. diam, by $3\frac{1}{2}$ ft. long. More coke is then added, and, when burning, is worked down with the poker. To do this the poker is thrust down lower than the muffle and is then carried through the coke with a motion tending to lift up the larger pieces while the smaller ones fall by and pack the space below the muffle. Fresh coke is now put in up to the level of the outlet flue and the crucibles are at once set, digging out a hole for them with the poker and tongs and setting them among the cold coke. With a good draft the coke is soon burning and the charge melting down. As soon as the muffle gets sufficiently hot, cupelling or scorifying can proceed. When the crucibles have been poured, the poker is again used to work down the coals, fresh coke is put in and a new batch of crucibles set. After their removal, poke down and replenish the fire before cupelling.

Even if there are no more crucible-melts to be made, still the fire must be poked down and replenished before cupelling.

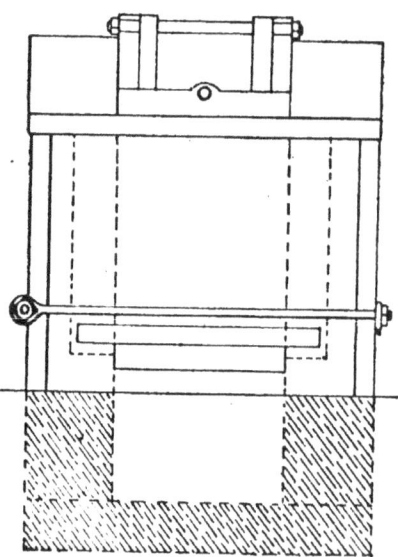

Fig. 16. Front Elevation of Wind-Furnace.

Fig. 15 represents a Bosworth furnace for a 9 by 15-in. muffle. consisting of a base, a body, and a top, on top of which may be noticed the outlet connecting to the chimney by means of a 5-in. stove-pipe. The top is covered by an iron door and, when the furnace is in full operation, it is filled with coke to the sill of the feed opening. The iron door covers the muffle, which latter extends clear to the back of the furnace and is capable of taking three rows of 20-gm. crucibles, two in a row, as seen in the cut. The plug immediately above the grate is kept closed except in cleaning out the furnace, and also when it is desired to poke the glowing coke and work it beneath the muffles.

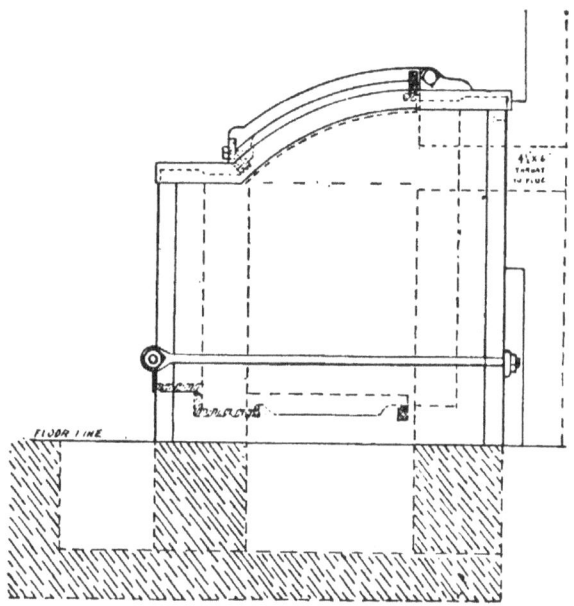

Fig. 17. Side Elevation of Wind-Furnace.

Fig. 16 and 17 represent a wind or natural-draft furnace intended for melting only. A thin fire or bed of coke is first started in the furnace. Fresh coke is added and the crucibles are at once set, being careful to place them to their full depth in the coke, and to pack it around them.

Gasoline Assay Furnace. Fig. 18 represents in perspective a combination gasoline furnace with burner at the rear, and the tank and pump by which the burner is fed. The furnace proper has at the front a muffle for cupelling, while at the top will be noticed an opening with the cover removed and showing crucibles in position for melting. The burner (Fig. 19) just fits into an opening through the rear wall. The lower hand-wheel admits a small quantity of gasoline to a tray shown below the burner, where it is ignited by a match. On opening the upper hand-wheel, the gasoline, under pressure from the tank which has been first of all pumped up, is heated and gasified as it enters the furnace from the jet. The issuing vapor is lit by a match imme-

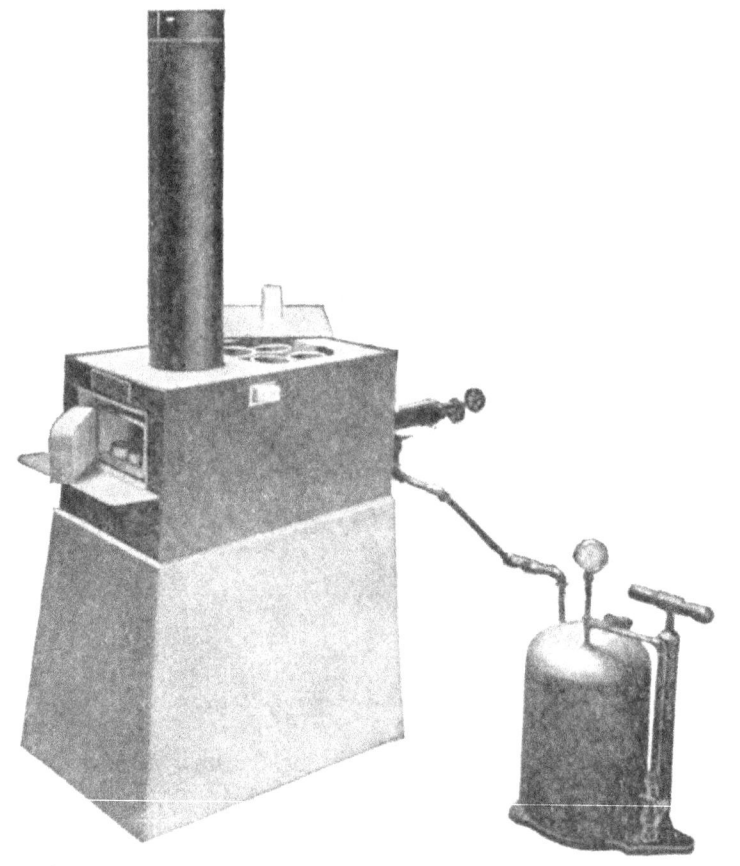

Fig. 18. Gasoline Assay Furnace with Tank and Burner.

Fig. 19. Burner for Gasoline Furnace.

diately inside the furnace, and as the latter heats, this gasification is kept up. For success in operating, therefore, it is necessary first to thoroughly heat the burner. The tank should be pumped to 10 or 15-lb. pressure. The crucibles to be melted may be placed within the furnace while it is cold and the heat brought up on them gradually. It is also possible to put them into a heated furnace, letting them take up the heat, and finally to start the burner. These furnaces have the advantage that they are quickly brought up to the full heat, which can then be retained indefinitely with but little further attention. They cost, for the furnace $25, for the burner $11, and for the pump, tank, pipe, and fittings, $20.

VI.—CRUCIBLES AND SCORIFIERS.

For ordinary assaying, but few out of the many kinds advertised are made use of. Thus, for work in the muffle, but two sizes are used, the 10 and the 20-gm. crucibles. (See Fig. 20.) These crucibles are short enough to be placed in the muffle of the coal-burning furnaces already described. Where the crucibles are to be set in the coke, the pointed form (Fig. 21) is used, sizes E and F.

FIG. 20. CLAY CRUCIBLES. (FLAT FORM.)

FIG. 21. CLAY CRUCIBLES. (POINTED FORM.)

Crucibles are often used but a single time. If still sound, as when assaying silicious ores, they may be again used. However, for ores or products containing but little of the desired metal, new ones should be employed. Besides the domestic crucibles, Battersea crucibles of English make are very durable. For lead assays, in order to clearly distinguish the crucibles from those used for a silver assay, the triangular crucible (Fig. 22) is used. These may be of the size 3.25 in. diam. by 3.5 in. high, called U size.

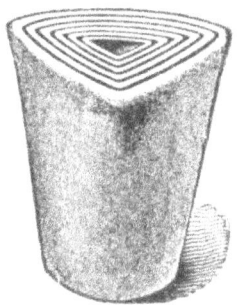

FIG. 22. NEST OF TRIANGULAR BATTERSEA CRUCIBLES.

Crucibles are made of fire-clay, white and smooth, or of a similar clay into which has been mixed a good deal of sand, so that they are rougher on the surface and of light-brown color. The latter better resist the action of a silicious charge, the former, or clay crucible, the basic charge.

In scorifying, the 2½-in. (outer diameter) scorifier, as shown in Fig. 23, is employed. They are good generally for a single

FIG. 23. SCORIFIERS.

time except for a silicious ore, and, even at the first time, may be so corroded as to permit the escape of a charge. The size specified is large enough for assaying one-tenth of an assay ton of ore, and, when larger quantities are to be taken, the crucible assay is chosen.

VII.—ASSAY BALANCES.

There are three kinds of balances used in assaying—the pulp, button, and gold-balances. In using any of these balances the variable load is placed in the right-hand pan, the fixed quantity in the left-hand one. Thus ore or silver granulations would be placed in the right-hand pan and buttons or fixed weights in the left. The beam is raised to its full height with a sudden movement, and, as soon as the pointer begins to move, the assayer judges what change is to be made, and at once drops the beam while he makes the needed change. Only once does he wait for the double swing, and that is when he is satisfied that the weighing is complete. In this way great rapidity can be obtained in weighing.

Fig. 24 shows a pulp-balance, sensitive to a single milligram, which, with a 2.5-in. pan, will take readily an assay-ton of ore. It is used for weighing ore, or the lead buttons of a lead assay. Price $15.

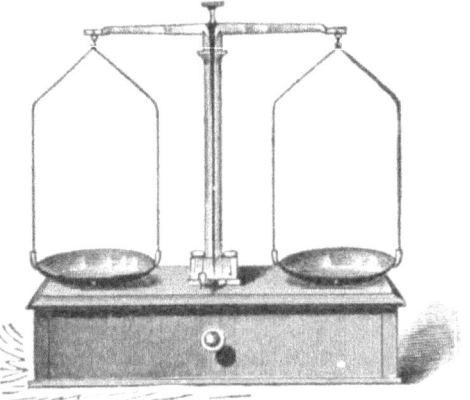

Fig. 24. Pulp Balance without Glass Case.

Fig. 25 shows a pulp-balance, enclosed in a case to protect it from air-currents, and sensitive to 0.25 mg. It can therefore be used, not only for weighing pulp and lead buttons, but even for chemical determinations, doing away with the necessity of purchasing a chemical balance where one wishes to economize in first cost of equipment. Price $30.

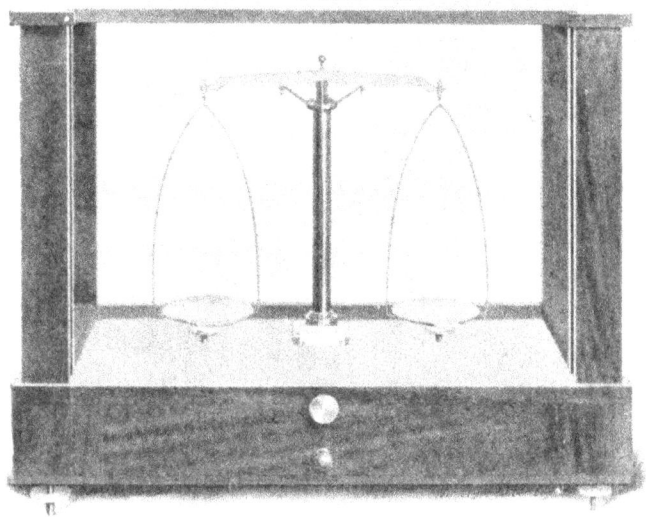

Fig. 25. Pulp Balance in a Glass Case.

Fig. 26 represents a so-called button-balance for weighing the small silver buttons obtained in assaying. It has a 6-in. beam, is sensitive to 1-50 mg., the beam being graduated to 5 mg. for a rider of that weight. In the drawers may be kept, on one side the pliers and button-brush, on the other, sheet-lead, copper, and silver-foil for the assay of fine silver bullion. The weights (500 mg. down) are conveniently kept on a neatly marked card-board or on a wooden strip, the weights being returned to their compartments at once from the pan. Never get into the way of displacing them. In handling the weights, the fingers must rest upon the case, and in moving the rider, the fingers must rest against the case to prevent sudden movement. Otherwise the weights are speedily bent and injured, and the rider may be thrown off the beam. Never leave the sliding door of the case open. It should be closed before you begin to use the rider. The balance can be leveled by means of adjusting screws, watching the bubble-tubes in the case. The final adjustment is made by means of a star-wheel or nut at the middle of the beam. It should be set so that the vibrations are equal on either side of the zero mark. Carry inside the balance a camel's-hair brush with which to brush off the pans before making the final adjustment. Sometimes, when the beam seems out of adjustment, it may be no more than a speck of dust in one of the pans, or even on the beam, itself. Price $110.

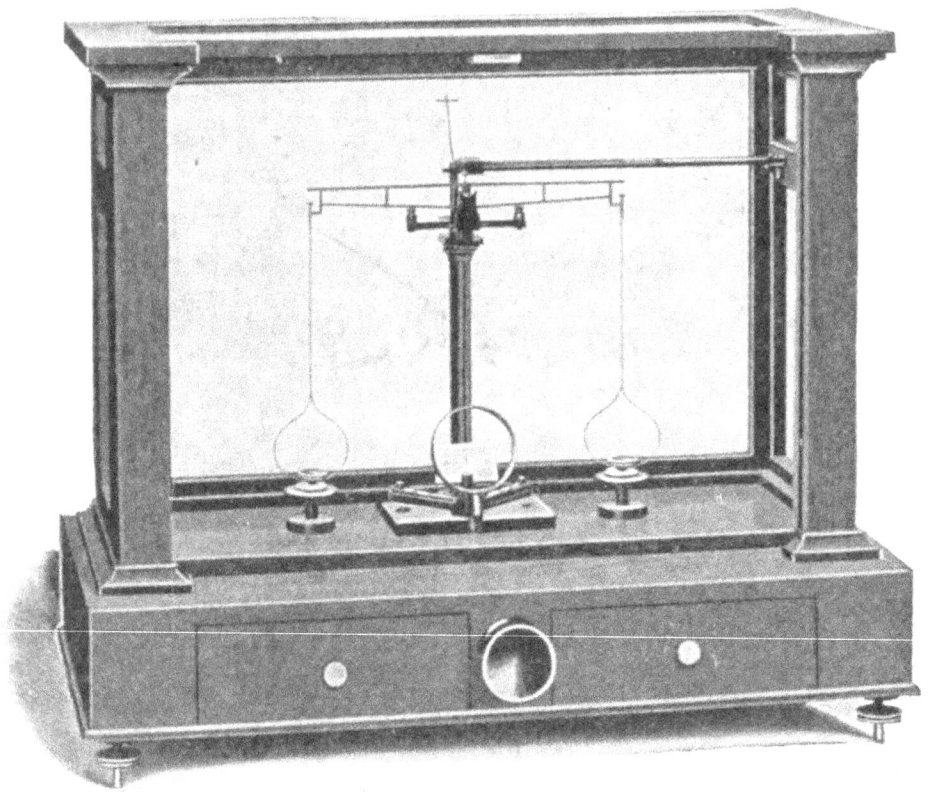

Fig. 26. Button Balance.

Fig. 27 represents an unusually fine balance for weighing the small particles of gold which are left upon parting the button of a silver-gold ore. It is sensitive to 1-200 mg. and has a 4-in. beam graduated for a 1-mg. rider. The knife-edges are of agate. For ease in reading the rider-divisions, pins project on the beam to direct the eye to the divisions. A rider may be carried on either side of the beam, though, since it can travel half the length of the beam, and read from the middle to the end, such an addition is hardly needed. The pointer divisions are viewed through a reading-glass. When the beam is dropped in weighing there is no kick or sudden movement of the pointer, and its direction of movement at once indicates the required adjustment. The beam is light and only 4 in. long, consequently the vibrations are rapid, and weighing is quickly concluded. Such a balance must be carefully shielded from irregular heating, such as the rays of the sun, and even the incandescent globe should be suspended centrally above it. The assayer must not omit testing to note whether the balance is in perfect adjustment before weighing. It will help him in the matter of speed if he also knows the value (in tenths of a milligram) of a vibration of the pointer on the divisions of the scale. Price $200.

FIG. 27. FINE GOLD BALANCE.

WEIGHTS. The weights used in assaying are gram-weights and assay-ton weights.

For use at the pulp-scale, for weighing the buttons from the lead assay, a set of weights from 50 gm. to 10 mg., as shown in Fig. 28, is sufficient.

For use at the button or gold-balances, a set of weights from 500 to 1 mg., accurately made, is sufficient.

Assay-ton weights, from one assay-ton to one-twentieth of an assay-ton, are used for weighing out the powdered ore. The assay-ton equals 29.166 gram, and contains, consequently, as many milligrams as there are Troy ounces (by which silver and

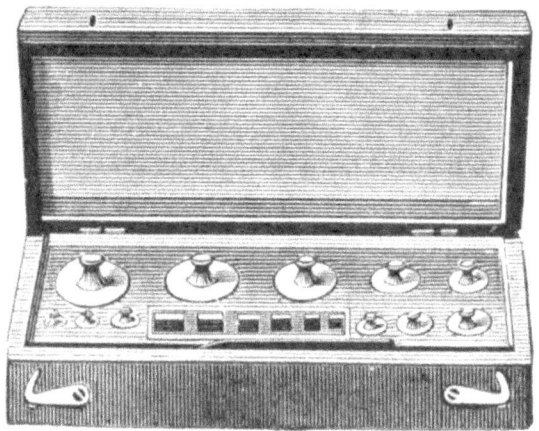

Fig. 28. Set of Gram Weights.

gold are weighed) in a ton avoirdupois of 2000 lb. Therefore, if one assay-ton (A. T.) of ore assays one milligram, the ton contains one ounce Troy. The value is thus obtained: A pound avoirdupois weighs 7000 grains and the Troy ounce 480 grains, therefore we have $\frac{7000}{480} = 14.58$ Troy ounces to the pound avoirdupois, or 2000 by 14.58 = 29,166 ounces per ton of 2000 pounds.

In practice, the weights of one gram and above are cylindrical and are handled with the fingers, those less than one gram are flat; and are never touched by the fingers but handled with pincers. Particular care must be taken of the finer weights, and, in placing them, the hand or fingers rest on the case. Where a number of weights are to be removed the pan is taken off with pincers and they are dumped from it upon the floor of the case and there counted. When returned to the box they are counted again as a check on the first reading. In weighing, proceed systematically. Thus if the weight is 378 mg., we would put on 500 mg., too much, then 200, too little, 200 more too much, 100 too little, 50 too little, 20 too little, 20 too much, 10 too little, 5 too little. The remainder we know to be between 375 and 380, which is to be determined by the rider which will weigh five milligrams on the beam.

The pincers or forceps for these weights are of brass with fine curved points, see Fig. 29, or nickel-plated as in Fig. 30. They

Fig. 29. Pincers.

Fig. 30. Nickel-Plated Pincers.

are used, not only for handling the weights, but also for picking up buttons, for granulations, or other fragments. In preparing the silver button for assay, it is removed from the cupel by means of pointed-nose button-pliers (Fig. 31), and the adherent bone-ash removed with the button brush (Fig. 32). For larger buttons, of over 100 mg., stouter pliers are to be preferred, since

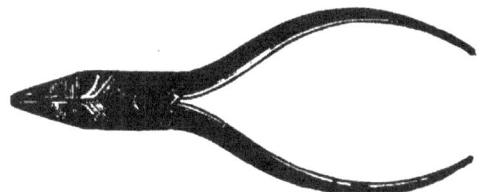

Fig. 31. Button Pliers.

Fig. 32. Button Brush.

the button must be gripped so tightly as to slightly deform it. Small buttons may be removed from the cupel with pincers, and placed on a smooth-faced steel block (Fig. 33) to be flattened with a light hammer (Fig. 34) as in blow-pipe work. The block is placed close to the pan in the balance and the button transferred from it directly to the pan. In this way one may lessen the chance of accidentally losing the button.

Fig. 33. Steel Anvil Block.

Fig. 34. Blow-Pipe Hammer.

For handling the lead button, when hammering it into shape and removing the attached slag, plain steel forceps are used, about 6 in. long. The hammering of the button should be upon a smooth-faced steel anvil set into a block (Fig. 35) weighing 4 to 6 lb., and with a smooth-faced hammer (Fig. 36) reserved for that purpose only, and weighing at least one pound.

Fig. 35. Steel Anvil to Set in Wooden Block.

Fig. 36. Slagging Hammer.

SPATULA. For weighing out and mixing ore, a spatula (Fig. 37) with a 6-in. blade serves very well. For weighing, an ordinary tinned iron teaspoon is even better, being cleaner and quicker. Indeed there are assayers who use this also for measuring out their fluxes.

MOISTURE SCALES. For determining the percentage of moisture in an ore, or the weight of fines and metallics in a sample, the moisture scale (Fig. 38) is well suited. It is finished with a

FIG. 37. SPATULA.

sliding weight, and weighs to 1 kg. A scoop can be placed on one pan and its counterpoise on the other, while extra weights beyond 1 kg. can be added.

FIG. 38. MOISTURE SCALES.

VIII.—FLUXES USED IN ASSAYING.

At the temperature of the assay furnace most ores are infusible, and in order to make them fusible, and to bring them into a uniform liquid mass in which the precious metals may come in contact with the collecting agent (molten lead) and be taken up by it, fluxes must be added. These fluxes, where basic, as in the case with soda, will form a fusible slag with silica as a silicate of soda; where acid, they form a fusible slag with the bases, as when borax is added. In addition, some carbonaceous substance, as argols or flour, will reduce litharge to metallic form.

We may divide fluxes into:
ACID FLUXES, as borax, glass, or sand.
BASIC FLUXES, as soda, pearlash, and litharge.
COLLECTING AGENTS, as litharge or test lead.
REDUCING AGENTS, as argols, flour, or coke dust.
OXIDIZING AGENTS, as nitre.
COVER, as salt.

BORAX. The ordinary crude borax of commerce is the prismatic neutral variety ($Na_2B_4O_7 + 10\ H_2O$) and when pure consists of:

Boracic acid (B_2O_6) 36.5%
Soda (Na_2O) 16.4%
Water of crystallization 47.1%
                                              ———
                                              100.0%

Borax-glass may be made by melting this crude borax in a crucible, or in an iron dish or ladle, taking care to fill the vessel only partly, since the hydrated salt swells to twice or more of its original bulk. Under the action of heat the water of composition is driven off, leaving white and opaque crystals, which melt down into a glass. More borax is added as the former portions melt down until the vessel is two-thirds full. The molten mass is now poured into a mold, or preferably upon an iron plate. The discoloration, due to iron, where an iron dish has been used, is a matter of indifference, the amount so dissolved being trifling. When cold, the borax-glass is broken in an iron mortar, and then ground upon the bucking plate until it will pass a flour-screen of 16-mesh. The assayer generally gets it already prepared for use.

For crucible assays, crude borax may be used. Placed upon the top of the charge, it tends to keep it cool to the last, and until the lower portion of the charge is melted, and tends to prevent the mechanical loss due to rapid escape of the gases. There is no danger to be feared from not mixing it with the rest of the charge, the melting proceeding quite as well. Where borax-glass is to be used it may be (and often is) mixed with the other fluxes and the ore. Borax-glass is most appropriately used in scorification, where it is placed upon the top of the charge in the scorifier.

Borax serves as an acid flux to unite with bases contained in the ore, and the quantity to be used depends upon the amount of bases present, as may be estimated either from the appearance of the ore, or better, by actual determination. The common bases are the oxides of the metals as, FeO, ZnO, MnO, CuO, and the alkaline earths CaO, MgO, and BaO. To this may be added Al_2O_3, which is, however, such a feeble base that it is quite commonly disregarded. The alkalies and other oxides, not above enumerated, occur to so small an extent that they are not taken into account. Alkalies occurring in feldspar, would, in a crucible assay, cut very little figure compared with the soda added as a flux. Where the amount of bases can be determined, it will be found, that for each part of base, we should use crude borax, four parts, and borax-glass, two parts. When so used it makes a fluid easy-pouring slag, and takes up and dissolves the oxidized bases. Such a slag is glassy and brittle. A deficiency of borax is indicated by the slag pouring too stiff or thick, but brittle.

Sometimes it will not leave the crucible, or else the pour is lumpy. However, a slag lacking soda may also act in this way, except that the silicious slag will string out and the trouble is to be laid to a shortage in borax, where we know that the amount of soda is right. Where a salt cover is used, and crude borax is employed, the latter is placed above the cover. It intumesces or swells up, and here, when much has to be added, is disposed to overflow before it again fuses and recedes into the crucible. If it appears disposed to thus go beyond bounds it may be pushed back into the crucible in such a way as to melt down. A small loss of borax is a matter of little moment. Crude borax loses half its weight by fusion to borax-glass, and hence twice the weight is needed to perform the same work as the borax-glass, as has already been indicated.

GLASS. Glass is a fusible silicate which must be free from lead when used for a lead assay. It contains 65 to 73% silica, and, being fusible (having been fused in the making) is preferred by many to sand for furnishing silica to a charge. Glass (besides silica) contains in general 20% of soda (Na_2O) or of potash (K_2O) together with lime (CaO). Lime, when present, is thought to give a cleaner separation of the button, and hence would be an advantage in assaying. Glass is pulverized to pass through a flour-screen. About one-third of the silica in glass is available as silica, the rest having been satisfied by the alkaline bases. Hence, in using it, three grams should be taken as equivalent to one gram of silica or sand.

SAND. A clean white sand, ground and sifted through a 60-mesh sieve, will serve to furnish silica to an assay-charge when needed. Both sand and glass protect the crucible from being eaten out under the action of basic fluxes (soda and litharge). We may reckon the whole weight of a quartz sand as available silica, though it is a pure sand which contains more than 95% SiO_2.

Either glass or sand is omitted from the regular assay, since it is found that it will disturb the calculation for reducing power, giving a button of a different size.

Soda. While bicarbonate of soda ($NaHCO_3$) and even the ordinary soda crystals have been used in assaying, the best for such use is the mono-carbonate made by the ammonia process, 98% pure and of the composition Na_2CO_3. When chemically pure it contains 58.5% caustic soda (Na_2O). It forms with silica at a red heat, a fusible silicate thus

$$Na_2CO_3 + SiO_2 = Na_2SiO_3 + CO_2$$

the escaping CO_2 gas thoroughly agitating and mixing the melting contents of the crucible. From the formula we find by calculation that one part of silica needs 1.76 part of soda to form the mono-silicate.

Pearlash, or commercial potassium carbonate, contains 90% K_2CO_3 together with some sodium carbonate, potassium sulphate, and sodic chloride; so that we have about 60% of caustic potash (K_2O) available to combine with silica. When the salt is mixed with the sodium mono-carbonate we get an alkaline flux which is more fusible than either of the carbonates taken separately, and this is the principal advantage which may be expected from such a mixture. The disadvantage of using pearlash is the introduction of sulphates into the charge, so that many assayers use sodium mono-carbonate only. Pearlash is disposed to absorb moisture from the air and become deliquescent, or at any rate, lumpy. When used with soda it is mixed in equal parts, or better, in the ratio of 2.7 parts of soda to 2.3 parts of pearlash. Those who use it claim it gives more satisfactory results than soda only, and habitually use it in mixture.

Litharge (PbO) is not only a basic flux, but that part which is reduced to lead is also a collector of gold and silver. It has the composition 92.8% Pb and 7.2% O. Ordinary commercial litharge contains a small amount of silver, amounting to 0.75 to 1.25 oz. per ton. If an assayer is compelled to use this, he should, after mixing the whole lot together to make it uniform, determine the amount of silver present, and deduct this from the assay according to the amount of litharge used. To make such a determination take: 30 gm. litharge, 2½ gm. argols, and 5 gm. sand. Bring the charge to quiet fusion, which will take 25 minutes. The lead button, weighing 20 gm., is cupelled.

There is now no occasion to use ordinary litharge, since pure litharge is readily obtainable from dealers in assay supplies.

When unreduced, and entering the slag, litharge is a basic flux, acting upon the silica thus, $PbO + SiO_2 = Pb\, SiO_3$; but also exerting a solvent action upon the sulphides present. A portion of the litharge, in an ordinary crucible assay, is reduced for the purpose of affording a lead button, into which are collected the gold and silver of the ore. Before its reduction, the litharge has exerted its oxidizing effect, and the impurities, thus oxidized, are ready for solution in that which has remained unreduced. As prescribed in the regular assay, the amount of litharge is largely increased where the impurities arsenic, antimony, copper, or tellurium are large in amount. The reducing power of various sulphides, or the amount of litharge needed to oxidize them, is, according to Berthier, as follows:

Manganese sulphide, 28 parts lead or 30 parts litharge.
Iron sulphide (FeS) 28 " " " 30 " "
Iron pyrite (FeS_2).. 46.5 " " " 50 " "
Copper sulphide ... 23 " " " 25 " "
Copper pyrite ..... 32.5 " " " 35 " "
Antimony sulphide . 23 " " " 25 " "
Zinc sulphide ...... 23 " " " 25 " "
Arsenic sulphide ... 46.5 " " " 50 " "
Lead sulphide ..... 2 " " " 2 " "

Red lead (Pb_3O_4) may be used in assaying in place of litharge. It contains 90.5% Pb, and has, because of its greater percentage of oxygen, a slightly greater oxidizing power upon sulphides.

REDUCING AGENTS. The reducing action of these compounds, such as charcoal, flour, argols, as well as sugar and starch, is due to their contained carbon acting thus,

$$PbO + C = Pb + CO \text{ (or } 2\, PbO + C = 2\, Pb + CO_2)$$

the carbon monoxide escaping as a gas.

CHARCOAL. Well-charred wood-charcoal is brought to a red heat in the muffle; it is then withdrawn, the ash blown off, and the pieces broken small in an iron mortar. The pulverizing is finished upon the grinding plate to pass a 40-mesh sieve. One part of charcoal will reduce 28 parts of lead from litharge and it is one of the most powerful reducing agents known. Its disadvantage is its lightness, whereby it tends to float on the surface of the charge into which it has been mixed.

FLOUR. The flour of wheat or of rye may be used. Its fineness enables it to be mixed intimately with the charge, and it is preferred by many to argols. One part reduces about 12 parts of lead from litharge.

ARGOLS. Argols is used like flour, being mixed into the charge. One part of it will reduce about eight parts of lead from litharge. When pure it has the formula $KHC_4H_4O_6$. Crude argols contains, besides bitartrate of potash, tartrate of lime, and the amount used in an assay is so small that its fluxing power is not regarded. It is, however, a basic flux. It has an advantage that its color prevents mistakes in using.

REDUCING-FLUX MIXTURES. Reducing-flux mixtures consist of a mixture of any of the above reducing-fluxes with soda, pearlash, or the two together. The advantage of using such a mixture over that of the ingredients separately, is that because of its dilution, a measured quantity will insure an exact reduction. Two common reducing-flux mixtures are

80 parts soda to 20 parts argols, and
85 parts soda to 15 parts flour.

The mixture should be intimate and uniform and the determination of the reducing power of a measureful should be made. Pearlash may be substituted in part for soda, or the alkaline carbonates may be used in ratio of 2.7 of soda to 2.3 of pearlash (in ratio of their atomic weights).

TABLE OF REDUCING POWERS. One part of each of the following reducing agents reduces as indicated:

Charcoal ....28 parts lead.
Wheat flour ....12 " "
Argols .... 8 " "
Reducing-flux; 80 soda, 20 argols......1.6 " "
Reducing-flux; 85 soda, 15 flour.......1.8 " "

This reducing power varies, and in any case it is best to determine it when a fresh lot has been prepared as follows:

Mix in a new crucible 3 gm. reducing-flux, 40 gm. PbO, and 10 gm. soda. Melt quickly and pour, just before the time of quiet fusion, into a conical-hole mold and weigh the resultant button. These, and in fact any fluxes, should be passed through a flour-sieve. Where a reducing mixture is made, mixing should be so thorough as to make all parts appear quite uniform in color.

OXIDIZING AGENTS. The nitrates of potash (nitre) and of soda, especially the former, are used for oxidizing the sulphides of the charge, where the ore itself would reduce too large a lead button. Their oxidizing power is such that one part will reduce from litharge:

Nitrate of soda ($NaNO_3$) 4.9 parts of lead.
Nitrate of potash (KNO_3) 4.2 parts of lead.

This is for the pure salt. We reckon the oxidizing power of commercial nitre at 4 even. Neither of them contains water of composition.

To determine this oxidizing power prepare two crucibles as follows: Charge into each 40 gm. PbO, 10 gm. soda, and 3 gm. argols, and to one of them add 3 gm. nitre. The crucibles are melted quickly and without loss, poured, and the resultant buttons weighed. The difference between the weights, divided by 3, will express the oxidizing power of the salt. As a result of the oxidizing action of the nitrates, their acid portion is removed, leaving the alkaline base.

SALT. Salt is used as a cover to the charge in assaying, and for this purpose, table salt or any ground salt is well suited. It cannot be regarded as a flux, and when melted, floats as a transparent fluid upon the charge. It then slowly volatilizes, giving a white fume. Where the charge is kept for a long time in the fire, it may be quite driven off. It will be found as a white layer upon the top of the poured slag. It acts as a wash, carrying down any particles which cling to the sides of the crucible. The commercial article contains some moisture, which decrepitates on heating. This may be removed by moderately heating the salt in an iron dish before using, where mechanical loss, due to decrepitation, is feared.

Many assayers have given up the use of salt, claiming that equally good results are attainable without it, and that it increases the chance of loss, especially where much nitre is used. Where salt is omitted, dependence is placed upon borax as a cover.

EFFECT OF EXCESSIVE FLUXING. It is desirable to use no more fluxes than necessary, since as the quantity of slag is increased, so is the silver and lead therein, and no slag is entirely clean. Where the fluxes are in excess the slag is fluid and thin, and is not as clean as the normal slag, especially in the lead assay.

IX.—ORES.

In order to determine what fluxes to use, the assayer must be acquainted with ores from their physical appearance.

They may be divided into oxidized and sulphide ores.

Oxidized ores, among which may be classed the carbonates, have a dull earthy look, are whitish or yellowish in color, and, when containing iron, brown and red. The depth of color is intensified where manganese is present, since manganese oxides tend to blacken the ore. At the same time, ores, apparently oxidized, may have some sulphides mixed with them and, on close inspection, the particles of zinc, iron, or lead sulphides may be detected. A preliminary determination for reducing power also indicates the same fact. Where there are shining particles of mica, do not confound them with the sulphides. Ores containing sulphur, as sulphate, are not, however, reducing in their action.

Sulphide ores have a semi-metallic appearance as shown in the case of galena, blende, and pyrite. Let us take the case of the three common sulphides, we have:

Galena containing 87% lead, 13% sulphur.
Blende containing 67% zinc, 33% sulphur.
Pyrite containing 46% iron, 54% sulphur.

It will be noticed that the percentage of metal (or base) decreases while the sulphur increases in this list. Since the lead in the galena is reduced to metal in assaying, it does not enter the slag, and, if so, from the point of view of the assayer, is not to be regarded as a base. Galena contains but little sulphur as compared with the other sulphides. Of the remaining sulphides (blende and pyrite), we may say that they contain much base, and also much sulphur.

Ores may be divided into silicious ores (having a gritty feel) and non-silicious ores, or quartz at one extreme, and galena or limestone at the other. Generally the non-silicious ores are also basic, as lime rock or iron ore, but galena or lead carbonate is non-silicious but not basic for the reason above given.

Again we may divide ores into silicious and basic ores, since the one quality is just the opposite of the other. An example of an oxidized basic ore is iron ore, and, among the sulphides, blende and pyrite.

As regards ores high in lead (leady ores) they are low both in silica and bases. Such ores feel heavy, and, when a portion is weighed out for assay, it occupies but little bulk. Thus the assayer is warned of its nature, and, in case of a lead carbonate, he can judge of the irony base it contains by the color. When, however, the sample has been dried at a rather high temperature this red color is intensified, and this should be taken into account.

In theory, the ore should be dried at a temperature not to exceed 100° C., since at higher temperatures sulphur on carbon dioxide may be driven off and the sample is in an unstable condition.

X.—THE SCORIFICATION ASSAY.

This is the simplest method of assaying, and is suited to all kinds of ores which come to the assayer. It is used for the determination of gold and silver in ores.

In making a scorification assay the following is the regular charge:

> 30 to 60 gm. test-lead, 0.1 assay-ton of the ore and 1 to 5 gm. borax.

In a 2½-in. scorifier measure one half the specified amount of test-lead, weigh out the ore, and mix with a spatula in the scorifier. Sprinkle on the remainder of the test-lead, completely covering the mixed portion, then add the required amount of borax. The quantity of borax will depend upon the amount of bases in the ore. When there is much base, add much borax, and for little base use little borax. For example, the pyrite or the blende ores, already mentioned, need 4 gm. borax, while a silicious non-basic ore needs but 1 gm., and that only as a safeguard.

The scorifier, thus prepared, and marked with red chalk or riddle to distinguish it from other assays to be done at the same time, is placed at the hottest part of the muffle, and the door of the muffle closed. The charge soon melts down, and, when in complete fusion, the door is opened and the scorifier brought toward the front of the muffle to scorify or 'drive.' Here it is exposed to the air, and the molten lead, at that high heat, begins to oxidize, giving off fumes. As the operation proceeds litharge forms, and, together with the borax, slags with the gangue of the ore. The slag accumulates upon the borders of the scorifier, and finally there is left exposed to the action of the air the middle of the bath only. This centre or eye diminishes to, say, ⅜ in. diam., when this stage is complete. The scorifier is again moved back to the hot portion of the muffle and the door closed. As soon as the slag is melted, the contents of the scorifier is poured into a slag mold. The scorifier, if the work is well performed, should be quite free from any undecomposed ore and smooth upon its whole interior surface.

It will be noticed that the scorification proceeds by three stages:
1. The ore is melted down (about five minutes).
2. The muffle is opened and scorification proceeds (about twenty minutes).
3. The scorifier is again heated (about five minutes) and is ready for pouring.

To return to the slag left cooling in the mold. When the contents of the mold is turned out, there will be found a lead-button, preferably of 10 or 12 gm., at the point of the mold, and above it the adhering slag. This slag is removed by hammering the button as it rests upon an anvil, so that the slag is thoroughly cleaned off and the button is ready for cupelling. It is to be understood that this button has gathered up all the gold and silver which was in the ore. This lead is removed in the operation of cupelling, and there remains behind a bead or button of silver (and gold).

CHEMICAL REACTIONS IN SCORIFICATION. In the first stage, as the charge (containing not quite three grams of ore) melts down, air comes in contact with it, tending to oxidize or roast any sulphides present according to the reaction:

$$PbS + 3O = PbO + SO_2$$
$$ZnS + 3O = ZnO + SO_2$$
$$FeS_2 + heat = FeS + S$$

The sulphur, on contact with air, burning to SO_2. Then:
$$FeS + 3O = FeO + SO_2$$

The molten lead is also oxidized to litharge as follows:
$$Pb + O = PbO$$

But any PbO, thus forming, eagerly takes up any silica in the gangue of the ore. The borax is busy dissolving the oxides as they form thus:

$$FeO + Na_2B_4O_7 = Na_2FeB_4O_8$$
$$ZnO + Na_2BO_4O_7 = Na_2ZnB_4O_8$$

By the time the door has been opened, enough PbO has been formed to dissolve the silicious gangue, and if not, the increased access of air at this time speedily produces it. At the period of driving, oxidation of litharge proceeds, which, reacting upon the partly decomposed sulphides, completes their roasting, thus:

$$PbS + 2PbO = 3Pb + SO_2$$
$$ZnS + 3PbO = 3Pb + ZnO + SO_2$$
$$FeS + 3PbO = 3Pb + FeO + SO_2$$
$$CuS + 3PbO = 3Pb + CuO + SO_2$$
$$As_2S_3 + 9PbO = As_2O_3 + 3SO_2 + 9Pb$$

That is to say, oxidation may be brought about by contact with the forming litharge as well as by the action of the air. As oxidation proceeds, the slag which is formed gradually covers the surface of the metal, leaving a continually smaller eye at the centre exposed to the air. When the eye at the centre becomes small, oxidation necessarily proceeds slowly; so that, to save time, it is better to pour the assay. The resultant button should weigh preferably 10 to 12 gm., although we often have it of 20 gm. weight, and sometimes no more than 5 grams.

GENERAL REMARKS ON SCORIFICATION. It will be noticed that the quantity of test-lead prescribed for an ordinary assay is 10 times the weight of the ore. When, however, the ore is 'dirty' (containing arsenic, antimony, copper, or tellurium) the lead is increased proportionately in accordance with the following table:

TABLE SHOWING THE QUALITY OF TEST-LEAD NEEDED PER $1/_{10}$ A. T. OF MINERAL.

MINERAL	GRAMS TEST LEAD	REMARKS
Antimonial ore	45	
Arsenical ore	45	Needs high heat in scorifying
Milling silver ore	30	
Cupriferous ore	30 to 60	
Zinc-bearing ore	30 to 45	Needs high heat and much borax
Leady ore	30	
Silicious ore	30	
Irony ore	30	
Telluride ore	30 to 60	

In scorifying, the air, as it passes over the contents of the scorifier, oxidizes the contained molten lead to litharge. The litharge reacts on the impurities of the ore, oxidizing them, and they enter the slag (formed by the litharge uniting itself to the gangue of the ore and to whatever borax has been added). The ore also tends to float on the surface of the lead, and is thus rapidly roasted in contact with the air. Borax is added to the charge according to the quantity of base present. Much base needs much borax. Thus iron ore, limestone, pyrite, and blende need much borax (especially the last, which tends to form a stiff slag). On the contrary, a silicious ore, a lead carbonate, or galena, needs but little. The base most commonly present is iron, and, in case of oxidized ore, the color gives some idea of its quantity.

XI.—CUPELLING.

PREPARATION OF THE CUPEL. Ground bone-ash is mixed in a pan with a little water until there is a coherence of the material when grasped in the hand. This is determined by experience. The bone-ash is placed in the mold, and the plunger brought squarely upon it. A blow or two of the mallet compresses the bone-ash in the ring. The cupel is ejected and placed upon a board, and the accumulated supply of cupels is placed in a warm place to dry. They can be used after thorough drying.

OPERATION OF CUPELLING. The cupel is warmed upon some portion of the furnace, is freed from dust by blowing upon it, and is then put into the hot part of the muffle. After remaining there until it has attained the heat of the muffle, the assay-piece is placed in it by means of cupel tongs. These must grip the cupel so as not to injure its top edges. The door of the muffle is closed, and the lead brought to its full heat until it clears. This is due to the melting of film of oxide which covers the assay-piece and which goes to the edge of the cupel leaving the surface exposed to the oxidizing influence of the air. The door is opened and the cupel brought further forward toward the front of the muffle to 'drive.' When the muffle is not so hot as to start this operation readily, it may be done by putting in a small piece of wood the size of the finger, which, laid in front of the cupel on the floor of the muffle, burns, producing a reducing atmosphere, while, at the same time, the door of the muffle is closed. This reduces the lead film already mentioned, so that when the door is opened, and the wood removed, cupellation at once proceeds. The temperature should be controlled between rather narrow limits. If it is permitted to become too low there is a liability of the button 'freezing' or solidifying and the assay is ruined. Even if the work can be pushed to the hottest part of the muffle and the assay started again, results will come out low, so that a button once frozen, especially toward the last of a cupellation, is useless. If the temperature is too high there will be a volatilization loss of silver, and, even when the operation is well performed, this loss will be as much as 2.5%. The proper temperature can only be learned by experience. The rule should be that the heat must be such that a slight ring of litharge crystals (feather litharge) forms around the sides, and especially at the front interior side of the cupel while the litharge-smoke is seen rising, and the top surface of the molten metal shows a slightly darker ring around its borders. As cupellation proceeds, the lead is oxidized and at once mostly absorbed by the bone-ash of the cupel, while a little passes off as a litharge-smoke. The button becomes gradually smaller and, at the expiration of perhaps 20 minutes, begins to round up, while, if the button is large enough, a play of colors may be seen passing over it as the last traces of lead are removed. At this time the temperature should be increased by pushing the cupel back to the hotter portion of the muffle. When the lead is all gone the play of colors ceases, the button brightens or 'blicks' suddenly, and the cupellation is complete. The cupel may be left for a little while longer in the muffle, since there can now be no further loss of silver. It is then removed, placed on a mold or an iron plate, and allowed to cool previous to weighing. If the button is large it is better to cover with a hot cupel before removing from the muffle and to remove it gradually, since otherwise it is liable to 'spit' or sprout, which may result in a loss through the projection of microscopic particles of silver from the substance of the button, which may be observed under the microscope at the borders of the cavity of the cupel. The reaction, $Pb + O = PbO = 51{,}000$ calories, indicates that much heat is evolved as the result of the burning of the lead to litharge.

The following table shows the losses which occur in cupelling a button of 10 gm. at different temperatures and the importance of performing this operation at the correct temperature:

Temperature Degrees C	Loss %	Remarks
700	1.05	Feather-litharge about button
775	1.18	Feather-litharge on cooler side
850	1.70	No more crystals
925	2.59	" " "
1000	4.78	" " "

XII.—PARTING.

The silver-gold button or bead, resulting from the cupellation, has now to be weighed upon the assay balance to the nearest tenth of a milligram, and the result at once written in the note book in ounces. (The weight in milligrams is not to be written in.) The button is taken from the cupel, being gripped strongly in the pliers (Fig. 32), and its lower surface brushed clean from bone-ash with a button-brush. The button is then placed on the table, the forceps seize it at right angles to the former position, and it is again brushed. Squeezing the button hard in the pliers should deform it somewhat, and thus tend to loosen any particles of bone-ash sticking to it.* The button may now be flattened with a hammer, after which it is ready for parting.

PARTING IN A PORCELAIN CAPSULE. A Royal-Berlin glazed porcelain capsule of 1⅜ in. diam. and 1¼ in. high, is preferably used in parting the button. Upon this is poured the pure dilute nitric acid (1 of acid of sp. gr. 1.42 to 3 of water), covering the button to say, ¼ in. depth. This is brought to the boiling point on the hot plate, taking care the ebullition is not violent, since such action tears apart the newly released gold. When action ceases, manipulate the particles of gold so as to bring them together. This can be effected with the aid of a glass rod, stirring the liquid in vortex fashion, and tapping the outside of the capsule to work the particles together. If a small particle of gold shows itself, floating on the surface of the liquid, it may be made to sink by touching it with the glass rod. It is always necessary to watch that no such particles are overlooked. The dilute acid is poured off into a jar or wide-mouthed bottle, leaving the gold behind. A few drops of stronger acid (7 of acid to 4 of water) are then added to the capsule, brought to boiling, until

*Let the beginner weigh the button without completely cleaning it, and then, after proper cleaning, weigh again. The difference in weight will be observable.

action of the gold has ceased, and then decanted. The capsule is now nearly filled with hot distilled water from a wash-bottle, directing the stream tangentially on the border of the capsule so as to disturb the gold as little as possible. This wash water is now poured off and the operation repeated.

Where distilled water cannot be obtained, the common water will form a precipitate of silver chloride, and, when this happens with the first addition of water, a few drops of ammonia, also added, will dissolve the chloride and clear the solution. Where common water has to be used to dilute the nitric acid, a small button of silver placed in the acid diluted with it will precipitate all chlorides present. This precipitate is filtered off before using the dilute acid.

The capsule is dried on a hot-plate, and ignited at a visible-red heat in the muffle. It is then removed, cooled by setting on a cold iron surface, and is ready for weighing. In weighing the particle of gold, first see that the balance is exactly adjusted. Wipe any particles of bone-ash from the exterior of the capsule, loosen the gold in it with the point of the pincers, and tap the capsule with the pincers so as to cause the particle to enter the pan, which latter has been removed and placed on the clean floor of the balance in readiness to do this. Weighing the gold is done to the one-hundredth of a milligram. A camel's-hair pencil brush may sometimes be used to advantage for removing final traces of gold. The results are registered thus:

1900			
Jan. 5	Ontario No. 27 (Irony ore.)		
	Ag and Au	15.8	
	Au	3.10	3.10
	Ag		12.7
	Pb 17.2		

PARTING WITH A FLASK. The button is dropped into a parting-flask, preferably one of 1-oz. capacity (See Fig. 39), which can stand on the hot-plate. Acid in moderate quantity, say to one-quarter inch in depth, is added, and the acid brought to the boiling point on the hot-plate. The action should be gentle, to avoid mechanical loss, and should go on until the nitrous acid fumes are driven from the solution. The acid is poured off,

FIG. 39. PARTING-FLASK.

and some of the stronger acid added in order to remove the last traces of silver. This is in turn brought to the boiling point and then poured off. Note, that in all cases the assayer is to see that the acid has completed its work before pouring it from the flask. Turn the flask so that the gold adheres to the upper side of it, and carefully pour in hot water, or add it from the wash-bottle, to half fill the flask. Turn back the flask to bring the gold beneath the water. Be careful not to break up the gold by violent action of the water. This wash-water having been removed, the flask is completely filled, a dry-cup of white clay, $1\frac{1}{4}$ in. diam., is inverted on the flask and the whole suddenly reversed. The gold is seen to fall through the water to the bottom of the cup. Cant it a little to one side, and quickly and quietly withdraw the flask, letting the water run out of it into a convenient jar. The gold in the cup is tapped together, if necessary, and the water poured from it. This work should be done promptly, or otherwise the dry-cup becomes soaked with water, and, when pouring off, the water will not be absorbed from the gold. The cup is dried, and ignited at a low-red heat, and the gold, now showing its characteristic yellow color, is weighed.

Any carelessness in parting may result in a loss of particles of gold giving a low result. Or again particles of dust may get into the parting capsule and if weighed up with the gold would give a high result. Where there are small amounts of gold, as two or three hundredths of a milligram, the assayer should adjust his balance just prior to weighing. He should never report small particles as a 'trace' until, after weighing, he has found that there is actually no change in weight.

Where the amount of gold in the button is high compared with silver, care must be taken that the silver is completely dissolved, using the stronger acid after the weaker, otherwise results will be reported too high. If you find upon careful treatment that not less than one-fourth of the original button is gold, you may be pretty sure that you have not dissolved out all the silver. In such a case it is necessary to add to this button, at least three times its weight of silver, wrapped in a small piece of lead-foil, and cupel off the lead in a small cupel. Or the same operation may be performed on charcoal, using but little lead-foil. The remaining button can then be parted properly and completely.

Right here the young assayer must be cautioned not to commit the blunder of reporting as gold a residue which may contain silver. No one will pay $20 per ounce for that which is worth but 70c, and an assayer who reports in this way is liable (and rightly) to lose his job.

XIII.—THE CRUCIBLE ASSAY.

While any ore may be assayed by scorification, the amount in which the determination is made is but one-tenth of an assay-ton, and consequently, whatever the error is in weighing, it is multiplied by ten in reporting the ounces per ton. Any variation in the sample is also exaggerated in the same way. In the crucible assay, however, one-half an assay-ton is commonly used, and results become more accurate. The difficulty may also be overcome by making several scorifications and uniting the silver buttons into one for the determination of the gold. This implies, however, a good deal of extra work, and the same result is attained by a single crucible assay.

Assays by crucible are made by one of two methods. The first is called the 'nail assay,' adapted to the treatment of clean ores, and the second, the 'regular assay,' also called litharge assay, intended for dirty ores, which carry impurities to be got rid of, such as copper, arsenic, antimony, or tellurium.

On the Pacific Coast as much as two to four assay-tons of ore are fused in large crucibles in a coke-furnace or in a crucible-furnace intended for melting purposes only. The method has the advantage that the determination is made in a large portion of ore, and consequently, on low-grade ores, with much accuracy.

THE NAIL ASSAY (for clean ores). The charge is as follows: 25 gm. soda, 20 gm. litharge, 3 or 4 gm. argols, 5 to 15 gm. borax-glass, 1 to 5 nails, and 0.5 assay-ton of the ore.

The various amounts are added by measure, there being a small cylindrical tin cup provided for each of the fluxes of the following dimensions:

For soda,	1½ in. diam.	by	1¾ in. high.
" litharge,	¾ " "	"	⅞ " "
" argols,	¾ " "	"	⅞ " "
" borax-glass,	1 " "	"	1⅛ " "
" test-lead,	¾ " "	"	⅜ " "

These measures can be made by any tinner. They must all be tried for their capacity when level full. In preparing the charge, the soda, litharge, and argols are measured at once into a 20-gm. crucible. The ore is then weighed out accurately upon the pulp-scales, and poured on top of these fluxes. A spatula with a 6-in. blade is now used to mix the contents of the crucible until they appear of a uniform color. The side of the crucible is tapped with the handle of the spatula to settle its contents, the borax is placed on top as a cover, and the nails are thrust into the charge, points downward.

The crucibles, thus prepared, may be placed at once in the hot muffle, though it is well to place them in the muffle when cold and allow the heat to come up gradually. Where the crucible is to be placed in the coke, fresh coke is added to the fire before so doing, so that the top part of the crucible is cold. This diminishes the chance of boiling over and consequent loss. The crucible remains in the fire until its contents are in tranquil fusion, the time varying, in general, from twenty minutes to half an hour. Sometimes, however, the time is greatly exceeded, as when an occasional bubble is seen rising through the molten charge. In any case the charge should be hot, well fused, and liquid, and should pour smoothly, without lumps and yet not be too thin or easily flowing. After pouring, the assayer should look into the crucible to see that it is perfectly smooth and with no unfused matter sticking to its sides. Sometimes scale from the nails, or a piece of one of them, may adhere to the crucible, but this can be scraped out with the cupel tongs. Upon removal of the crucible from the furnace, hold it by the tongs in the left hand, and, with the right, use a pair of short tongs or forceps (Fig. 4) to take out the nails. Now, still holding the crucible, tap it upon the table or plate where you are doing the pouring, and then neatly pour into a conical mold (Fig. 10). After pouring, do not disturb the mold for five minutes or more, or until the contents have cooled and set. If you have to move the mold, keep it perfectly level. Otherwise you have the lead forming a fin at the side of the mold. When set, the cone of slag and the button are slid out on the grinding plate, or on an anvil, the button is detached from the slag and hammered into the form of a cube. The object of the hammering is to remove the slag, and, when the button is thus cleaned, it is ready for the next operation of cupelling. Any slag left adherent to the button shows itself when cupelling as a slight crust at the edges of the cupel, and many retain a little lead, thus causing a loss.

Cupelling should be done at a low temperature so as to make feather-litharge upon the surface of the cupel, otherwise there is an increased cupellation-loss by volatilization. For gold, this is less important. Toward the end of the cupelling, and as the silver button begins to show up, the cupel should be shoved back an inch or two to where the muffle is hotter. This is to insure that the last traces of lead are removed. By close watching you may see the 'blick' or sudden brightening of the silver bead which remains upon the cupel. Some assayers think that at this stage it is essential to remove the cupel from the furnace, but, evidently, when the silver button has become solid, as it now does, no further silver loss can occur.

The silver bead, when brushed to free it from bone-ash, may be weighed, then flattened, dropped into hot dilute nitric acid in the parting capsule, and parted, as already described under the head of scorification. If, however, the amount of gold by weight approximates one-fourth of the button, this treatment, with weaker acid, must be followed by one with acid of 1.3 sp. gr. (7 parts strong acid to 4 parts water). When the gold is in excess of this specified amount, parting proceeds slowly and imperfectly, and the silver cannot all be removed. In such a case the button is wrapped up in two or three grams of lead foil together with a little silver foil, in quantity such that the total silver present shall be at least three times that of the contained gold. This is quickly cupelled on the old cupel or on another small one of one inch diameter. The resultant button can now be easily parted. Where it is not desired to determine the silver, but the gold only, a small piece of silver foil may be added to the lead button just before cupelling, or it may be added in the crucible.

ACTION OF THE FLUXES. As regards the action of the soda and borax, an easy aid to the memory is to say: S equals S and b equals b, or that the soda is intended to flux the silica and the borax the bases. In an extremely silicious ore the slag, even with a good heat, may be stiff, so that when poured it runs with difficulty (sometimes not at all) from the crucible. Now, the 25 gm. soda already prescribed is sufficient for all ores except the very silicious ones, and when, as in the above case, this trouble threatens, it may be got over by adding another 10 gm. soda direct to the crucible while still in the furnace. The addition of borax will not help. When grinding up the ore on the plate, its gritty feel is a sign that the ore is silicious. An ore consisting of limestone, dolomite, or heavy spar, while of a light color, would not have a gritty feel under the muller.

On the other hand an ore which the assayer knows to be basic, if acting in this way, would need more borax (but not more soda) added to the charge, in order to improve its fluidity.

The way the scarcely fused slag acts is significant. When it can be strung out in threads like molten glass, it is silicious, but when it breaks short, as though brittle, in a way to be learned by observation, it is too basic.

Referring to the table and the comments thereon (Chapter IX), it will be noticed, that in the nail-assay, the lead is not to be regarded as a base which will go into the slag, and hence no borax is to be provided for because of its presence. With zinc and iron present, as in blende or pyrite, in calamine or in iron ore, or with the alkaline earths as in heavy spar and limestone, it is a different matter, since these bases enter the slag and they need borax to flux them. The amount of borax prescribed is from 5 to 15 gm. for the half assay-ton of ore taken. The smaller amount of 5 gm. is no more than enough to form a cover to the charge, though enough for small amounts of bases. The maximum amount of 15 gm. is to be used where there is much base present. In particular, when there is much blende, we would use the maximum amount of borax, since blende is disposed to make a stiff slag.

In the nail-assay, enough litharge is used to form a lead button of 18 to 20 grams, which is large enough to remove or collect all the silver or gold which the ore contains. Since an excess of argols is also used, this lead is all reduced. Should there be lead in the ore, this would also go into the lead button, and hence the assayer must take care correspondingly to lessen the litharge. For example, in the case of a galena carrying say 70% lead there would be 10.5 gm. lead, and hence only half the usual quantity of litharge would be needed. There is considerable lee-way allowable in the size of the button. As little as 15 or as much as 25 gm. will equally serve, but when below the former amount, it is doubtful whether all the gold and silver is collected, while, where the button is so large as 25 gm., it is longer in cupelling, and volatilization loss increases.

Since the reducing power of a gram of argols is such as will reduce approximately 8 gm. lead, the amount of 3 or 4 gm. prescribed will be in excess of all requirements, so that all the lead present in the charge is bound to be reduced.

The limits in the number of 10-penny nails are from one to six, and the number is dependent upon the quantity of sulphur present. Where the ore consists of pyrite, or of blende, the larger number of nails is used, in the case of galena two, while those ores containing little or no sulphur need but a single nail. One nail at least is used, since it may be possible, even in a supposedly oxidized ore, that there is still a little sulphur. At any rate a nail can do no harm. It cannot be said that too many nails are an objection, except as they take time to remove from the crucible before pouring, and that a drop of lead from the button might become adherent to one of them.

To show how much borax and how many nails are needed in assaying sulphide ores, let us take the case of the following sulphides. They contain, when pure, as follows:

Galena	PbS............87%	Pb......13%	S
Blende	ZnS............67%	Zn......33%	S
Pyrite	FeS............47%	Fe......53%	S

It will be seen, that starting with blende, its sulphur is 33%. The zinc makes up the remaining 67% and is twice the sulphur, that is, in blende, the sulphur makes up one-third, the zinc two-thirds. Next we find the quantities of base varying by 20% up and down. Thus all the numbers are deduced if we know the sulphur in blende.

Now as to the action of borax. Referring to our data we need not consider the lead as a base, since it enters the button, not showing up in the slag at all. Of the two remaining bases the zinc is more basic, blende needing more borax than pyrite. For the sulphur contents we would conclude, that the pyrite and blende need the maximum amount of six nails, the galena two. As a matter of fact we use many for both blende and pyrite, and few for galena.

CHEMICAL REACTIONS DURING FUSION. Lead oxide is acted on by the carbon in the argols, the reaction being as follows:

$$PbO + C = Pb + CO$$
$$CO + PbO = CO_2 + Pb$$

The button, thus reduced, carries down with it the gold and silver contained in the ore. If copper and antimony are present, they are reduced also, making a hard button, if there is much of either metal. When such a lead button is cupelled, some gold or silver is carried into the cupel, so that the nail-assay is suited only to clean ores.

Silica, itself infusible at high temperatures, is fluxed by soda thus:

$$SiO_2 + Na_2CO_3 = Na_2SiO_3 + CO_2$$

the carbon dioxide passing off as a gas. By calculation, there must be 26 gm. sodium carbonate to form the mono-silicate where there is present 15 gm. (nearly one-half assay-ton) silica as in the case of a pure quartz ore.

If crude borax is used in the assay, its water of composition is first given off thus:

$$Na_2B_4O_7 + 10H_2O + heat = Na_2B_4O_7$$

At first it swells up, sometimes to the extent of overrunning the side of the crucible, then melts down into borax-glass, hence the advantage of using this latter. Borax has a solvent effect upon bases, much as follows:

$$FeO + Na_2B_4O_7 = Na_2FeB_4O_8$$
$$CaO + Na_2B_4O_7 = Na_2CaB_4O_8$$

forming fusible borates. Borax, however, dissolves bases in all proportions.

The nails act upon the sulphides much as follows:

$$FeS_2 + Fe = 2FeS$$

The iron sulphide or matte thus formed is dissolved by an alkaline carbonate, hence the necessity of having plenty of soda at all times in the charge. Galena is decomposed as follows:

$$PbS + Fe = Pb + FeS$$

and for lead in oxidized form

$$PbO + Fe = FeO + Pb$$

so that, even were there no argols present, the lead would be reduced to metallic form. Blende is decomposed according to the equation:

$$ZnS + Fe + PbO = FeS + ZnO + Pb$$

THE REGULAR OR LITHARGE ASSAY (also called 'nitre assay'). This is used for impure ores, or those containing certain impurities as: As, Sb, Fe, and Cu, which must be eliminated from the lead button before cupelling. To do this the following charge is made up:

25 gm. soda (mono-carbonate), 30 to 120 gm. litharge, 0.5 A. T. ore, 3 to 15 gm. borax-glass, with argols or nitre to suit.

The quantity, either of argols or nitre, which is to be used will depend upon the reducing power of the ore, and is so regulated that a lead button of 20 gm. is reduced from the litharge in the charge.

To ascertain this reducing power we take a preliminary charge as follows:

0.1 A. T. ore, 40 gm. litharge, and 10 gm. soda.

This is melted down rapidly in a 10-gm. crucible, withdrawing it, even before quiet fusion, and pouring into a conical mold. The lead button obtained will vary, from nothing if it is an oxidized ore, to upward of 30 gm. if the ore is an iron pyrite, the reducing action being due to the presence of sulphur in sulphides. If the reducing power of this one-tenth be multiplied by five we have that of a half assay-ton. We may have either:

1. The reducing power of one-tenth A. T. less than 4. In this case multiply by 5, subtract the product from 20, and divide the remainder by the reducing power of the argols, say 8. This will give us the number of grams of argols to be used. For example, let the reducing power be 2. Then

$$\frac{20 - (2 \times 5)}{8} = 1.25 \text{ gm.},$$

the quantity of argols to add in the regular assay.

2. In the second case, the reducing power being greater than 4, we multiply by 5, subtract 20 from it, and divide the remainder by 4, the oxidizing power of nitre. Thus suppose the reducing power to be 22; then we have

$$\frac{(22 \times 5) - 20}{4} = 22.5 \text{ gm.}$$

to be added to the charge in order to secure a button of 20 grams.

Returning to a consideration of the regular charge we may note:

The soda is added, as in the nail-assay, to flux the silica.

The litharge varies according to the impurities in the ore, which may be estimated from the appearance of the ore, or from experience with it. For an ore containing but little of them, we may use 30 gm., while for ores carrying much arsenic, antimony, tellurium, or copper, we use to the limit of 120 gm. PbO. It will be seen, that since but 20 gm. is reduced from the litharge, there will be left anywhere from 10 to 100 gm., which goes into the slag. It is this litharge which acts on these impurities, oxidizing and dissolving them.

The borax-glass acts precisely as in the nail-assay, forming double borates with the bases.

This charge is placed in a 20-gm. crucible, is mixed there with a spatula, and the borax sprinkled upon it to form a cover. It is placed, either in the muffle or among the coals of a wind-furnace, brought to fusion, but at such a rate as not to permit any of its contents to escape from the crucible or to boil over, and finally brought to quiet fusion. This work will take from one-half to one hour. There is no objection to retaining the melt in the furnace even after this time. On the other hand, if the crucible is withdrawn before reduction is complete, not all the silver is obtained. In pouring, rest the lip of the crucible upon the mold, and pour at an even rate. If the slag appears to be stiff, when melting down an ore known to be silicious, it may be made more fluid by the addition of soda. For a known basic ore, on the contrary, we must add borax to produce the same effect. This addition may be made by putting in the amount by means of a long-handled spoon forged on the end of an iron rod two feet long, or, in an off-hand way, by means of a scorifying dish held by the crucible tongs. The button which results from this assay should be soft, and when cupelled, show freedom from impurities. If not in this condition, the assay should be repeated, using a much larger quantity of litharge.

CHEMICAL REACTIONS OF THE REGULAR ASSAY. A sulphide ore, having a high reducing power, is permitted, in the regular assay, to act upon and reduce from litharge 20 gm. lead, much as follows:

$$FeS_2 + 5PbO = 5Pb + FeO + 2SO_2$$
$$PbS + 2PbO = 3Pb + SO_2$$
$$ZnS + 3PbO = 3Pb + SO_2 + ZnO$$

The sulphur dioxide escapes as a gas, and the oxides are dissolved by the borax. But there remains a further amount of the sulphides, for the oxidation of which nitre has been provided, and which, at a high temperature, is decomposed.

$$2KNO_3 = K_2O + 2NO + 3O$$

the oxygen thus furnished acting directly upon the sulphides as for example:

$$FeS_2 + 5O = FeO + 2SO_2$$
$$CuS + 3O = CuO + SO_2$$

Also

$$SO_2 + O = SO_3$$
$$K_2O + SO_3 = K_2SO_4$$

and this floats on top of the charge.

We may note here, that where, in the nail-assay, copper is reduced to metallic form entering the button, it is here oxidized and goes into the slag. Arsenic and antimony sulphides are also oxidized according to the formula:

$$As_2S_3 + 9O = As_2O_3 + 3SO_2$$
$$Sb_2S_3 + 9O = Sb_2O_3 + 3SO_2$$

and enter the slag, being dissolved by the litharge.

Oxidized ores have no reducing power, and to them we add argols, which act as in the nail-assay:

$$PbO + C = Pb + CO$$

We are careful in this case to add no more (as may be determined by a preliminary assay) than will reduce the 20-gm. button, since it is essential that there be litharge to enter the slag to dissolve impurities.

XIV.—ROASTING OF ORES.

Where an ore contains but little gold, and we wish, therefore, to determine it in a large quantity, the ore may be first roasted. To do this, weigh out one assay-ton of the ore and put it in a 4-in. roasting-dish. The dish is first carried at a low heat at the front of the muffle. Where there is much pyrite present it will snap and fly, causing mechanical loss. This may be guarded against by partly covering the dish with a crucible cover, leaving the front side of the dish exposed. The dish is gradually advanced into the muffle, finishing the ore at a red heat, but so that it is not fused together. During the roasting the contents of the dish are stirred several times to expose fresh surfaces to the action of the air. Roasting may be regarded as complete when no odor of sulphur is observable upon removing the dish and smelling of its contents.

CHEMISTRY OF ROASTING. The ore may consist of a silicious gangue together with sulphides of iron, copper, zinc, and lead. Of these, pyrite is the most easily decomposed, the first equivalent of sulphur being driven off at a low temperature thus:

$$FeS_2 + heat = FeS + S$$

The sulphur, as it is evolved, encounters the air and is burned to SO_2. With an abundance of air, and at an increased temperature, the iron sulphide is roasted as follows:

$$3FeS + 11O = 2SO_2 + Fe_2O_3 + FeSO_4$$

The odor from the escaping sulphuric dioxide is quite evident. We also have when there is chalcopyrite,

$$FeCuS_2 + 6O = CuO + FeO + 2SO_2$$

As the heat is increased to 590°, the iron sulphate becomes decomposed, transferring its SO_3 to the copper thus,

$$FeSO_4 + CuO = FeO + CuSO_4$$

The ferrous oxide is then further oxidized, giving, where much of it is present, a reddish color to the roast, namely,

$$2FeO + O = Fe_2O_3$$

Zinc and lead sulphides are roasted as follows:

$$ZnS + 3O = ZnO + SO_2$$
$$PbS + 3O = PbO + SO_2$$

and at a still higher temperature (750°) the copper sulphate is decomposed, transferring its sulphuric anyhydride to the zinc and lead compounds, and forming sulphates,

$$CuSO_4 + ZnO = ZnSO_4 + CuO$$
$$CuSO_4 + PbO = PbSO_4 + CuO$$

These newly formed sulphates are hard to decompose (above 1000° being needed) by the action of heat, especially since the finishing heat must be high, silver will be volatilized. When so roasted, these sulphates do not affect the reducing power of the ore. Antimony and especially arsenic are oxidized, forming volatile compounds, and are thus, in part at least, removed.

$$As_2S_3 + 9O = As_2O_3 + 3SO_2$$

Ores, because of the time and attention needed, are seldom roasted; the assayer generally takes the shorter method of a nail or regular assay, making them in duplicate, and combining the cupelled buttons when the amount of gold is but small.

THE ASSAY OF ROASTED ORE FOR SILVER AND GOLD. One assay-ton of ore has been roasted so that it is oxidized and has lost its reducing power. We accordingly take: 1 A. T. ore, 40 gm. soda, 40 gm. litharge, 20 gm. borax, and 2.5 gm. argols.

This is placed in a No. F crucible, if it is to be done in a wind-furnace, the borax being used as usual as a cover. The excess of litharge, over that needed to form the lead button, is intended to take care of the impurities present. Where there is much of these, the roasting method would not be followed.

The charge is melted down to quiet fusion, and is then poured into a conical mold. The quantity of slag may be so much as to overflow the mold, but this does not matter. A 20-gm. button should result. In other respects the assay is conducted as in the regular assay.

XV.—ASSAY OF MATTE.

This is an artificial iron-copper sulphide, a furnace product consisting of iron and sulphur, the iron being more or less replaced by copper and perhaps lead. It contains silver and gold taken from the ore-charge, and is assayed by scorification using 45 to 60 gm. test-lead, 0.1 A. T. the ore, and 2 to 3 gm. borax.

The higher the matte in copper, the more test-lead we add. In cupelling one can tell by the appearance of the cupel whether the copper has been entirely removed from the button, since at the finish the surface of the cupel immediately around the button is a little lighter in color than the rest. Also, upon dissolving the button, if the solution is neutralized with a little ammonia, the absence of blue color denotes freedom from copper.

XVI.—HIGH-GRADE SILVER-SULPHIDE ASSAY.

The assayer may be called upon to assay ores rich in silver, or silver sulphides of several thousand ounces value, the product of a silver-lixiviation process. The sulphides may contain much copper, which has been precipitated at the same time with the silver, so that we scorify the sulphide, making a charge as follows: 0.1 A. T. ore, 45 gm. test-lead, and 2 gm. borax-glass.

It will be noticed that the usual quantity of 30 gm. test-lead has been increased because of the presence of copper. A clean ore, no matter how rich, can be quite as well assayed with the usual 30 gm. The borax-glass is added to take care of the copper oxide.

When the button resulting from this scorification has been cupelled, the litharge-soaked part of the cupel is saved, together with the slag of the scorification. These two are then coarsely powdered in a mortar, or on the grinding plate, to about 20-mesh and assayed with the following charge: Slag and cupel, 30 gm. litharge, 20 gm. soda, 2.5 gm. argols, and 30 gm. borax.

The bone-ash of the cupel, being chiefly calcium phosphate, needs much borax to flux its lime base. The material, being quite oxidized, needs just enough argols to give the usual 20-gm. button. A small amount of silver is recovered, which is added to the other for the total gold-silver button, and the result in milligrams multiplied by 10. The buttons are then parted for gold. Generally four to six results are made from as many tenths of an assay-ton. The buttons may be parted two and two, so that the gold is weighed in one-fifth of an assay-ton, and the results must check one another.

XVII.—ASSAY OF CYANIDE SOLUTIONS.

Cyanide solutions, containing silver and gold, may be either the solution containing the precious metals before extraction, or the solution after it has passed through the zinc-boxes, having had its values mostly removed (barren solution). In the latter case a much larger portion of the solution must be taken for assay than in the former.

Of the numerous methods devised we give two.

FIRST METHOD. This is accurate and simple, but slower than the second. It consists in evaporating to dryness upon a hot-plate, say 2 A. T. solution in a lead-tray made by turning up the edges of lead-foil such as is used in assaying. This is formed over a wooden block and may be 2 by 2 by ¾ in. deep. The evaporation being finished the tray is then bent into a compact form and directly cupelled. If, however, the residue is considerable, it may be advisable to scorify the lead before cupelling. The results are to be reported in ounces per ton of solution.

SECOND METHOD. This method is more rapid than the first, and can be used with a larger bulk of solution, and is performed as follows:

Measure into a No. 4 beaker 5 A. T. or 146 c.c. of the solution. (If the solution is a 'barren one,' 10 A. T. should be used.) Add enough KCy to the solution to bring it up to 0.5%, KCy.* Thus if the strength is 0.2% we shall need to bring it to 0.5% the addition of 0.3% of 146 gm., or 0.43 gm. of the salt. Stir into the solution 10 c.c. of a 10% solution of lead acetate ($Pb\overline{A}$) slightly acidified with acetic acid. Add from the point of a spatula about 0.4 gm. zinc dust, weighing it until able to judge of the quantity needed. This amount of zinc is amply sufficient to precipitate not only all the lead, but also the gold and silver in the solution. Thus we have:

AUSTIN'S FIRE ASSAY.

$$Zn + Pb\overline{A} = Zn\overline{A} + Pb$$
$$Zn + 2KAuCy_2 = K_2ZnCy_4 + 2Au$$
$$Zn + 2KAgCy_2 = K_2ZnCy_4 + 2Ag$$

Stir it and bring to boiling. Add 15 c.c. strong HCl to dissolve the excess of zinc. Leave the solution on the hot plate until the zinc is completely dissolved, and the lead has gathered into a sponge. As soon as thus dissolved, remove from the heat to prevent the lead from being at all dissolved. Holding the lead with a flattened glass rod, decant the solution. Wash the sponge with water, squeeze it into a clot with the fingers, and place it upon a piece of lead-foil weighing 5 to 6 gm. Fold the foil so as to leave an outlet for steam, then cupel it. The resultant button will give the silver and gold in the 5 A. T. solution.

---

* To determine the strength of the KCy solution we may proceed as follows: Have 50 c.c. burettes. Fill one with $AgNO_3$ solution consisting of 17 gm. of the pure salt in one litre of water. Fill the other with the solution to be tested. Measure into a small beaker, 13 gm. of the KCy solution. Run in cautiously the standard $AgNO_3$ solution until the white precipitate, which has appeared, ceases to dissolve where the contents of the beaker has been stirred, that is, when only a faint opalescense remains. Ten c.c. of the standard solution equals 1%, so that the reading is to be divided by 10. Thus suppose our reading was 14.5 c.c., then we will have 1.45% KCy present. For very dilute solutions we may take 130 c.c., and then the burette reading must be divided by 100.

XVIII.—ASSAY OF BASE-BULLION.

Base-bullion, or work-lead, is the product of the silver-lead blast-furnace, and may be considered as an unrefined lead containing upward of one per cent of silver together with a few per cent, at most, of impurities, as arsenic, antimony, and copper.

The sample may come into the assayer in the form of cylindrical 'chips' 2 in. long by ⅛ in. diam. Since two of these pieces are taken from each ingot or bar, there would be 800 of them for a 400-bar carload, and amounting in quantity to a quart. A plumbago crucible is made red-hot in the wind-furnace, and the chips are shot into it by means of a scoop. The lead quickly melts down, and is stirred with a wooden stick. It is then lifted by

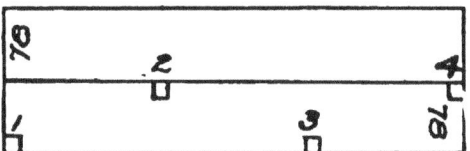

FIG. 40. BASE-BULLION SAMPLE.

the tongs, given a swirling motion to complete the mixing, and quickly poured into a mold 8 in. long by 2.5 in. wide, where it forms a bar about 0.3 to 0.4 in. thick. This bar is cut longitudinally, by means of a chisel, into two, each half being numbered; one of these is sent to the refinery and one retained by the works producing the base-bullion. Assuming that the lower half is retained, four pieces marked in as 1, 2, 3, and 4, each of a little more than 0.5 A. T. are cut from the bar. They represent a fair average of the bar, since the exterior runs a little higher than the central portion at the time it solidifies.

In case of an impure bullion, each sample should be scorified with extra test-lead and a little borax before cupelling. Where the sample is sufficiently pure, direct cupellation is satisfactory.

Some assayers prefer always to scorify before cupellation, contending that the loss is less in scorifying than in cupelling. Cupellation should proceed with the formation of feather-litharge, and, just prior to brightening, the cupel should be pushed back a little to the hotter part of the muffle. Where the buttons are large it is a good plan to have cupels set just behind in the muffle to cover them when finished as in the assay of silver bars. The buttons are parted as in ores.

XIX.—ASSAY OF SILVER BARS OR INGOTS.

Silver bars are assayed to determine their fineness, or parts in a thousand. Thus U. S. coin silver is 90% or 900 fine, the residue being copper. Ordinary ingots or bars from stamp-mills may vary from 250 to 900 fine in silver and gold, the impurity being copper. By fine silver we mean pure silver bars of 1000 fine, though the term is also applied to bars of over 900 fine.

Mill-bars may be assayed from granulations or from chips. To make granulations, when the silver has been melted and refined in a crucible, and just before it is poured into ingots, some of the molten metal is removed by means of a heated ladle or scorifier, and poured from a height of three feet into a bucketful of water. The metal is broken into fragments and shot of all sizes from which may be selected particles of any desired size. Where the ingots come to the assayer ready cast, he takes his sample by chips cut from opposite edges of the bar, the chisel cutting in a converging way into it to take out a wedge-shaped chip of from 2 to 5 gm. These are placed in a scorifying dish in the muffle, and brought to a just-visible red heat in order to anneal them. They are then hammered out upon an anvil, again annealed, and put through rolls (Fig. 41) where they are brought down thin enough to cut readily with scissors. Before use this sheet-metal is cleaned by rubbing with emery cloth.

The operation is as follows:

TRIAL ASSAY. To determine approximately the fineness of the sample, we weigh out 500 mg. upon a button-balance, selecting from the granulations larger pieces, and finally the finest particles, to obtain the exact amount. This weighing need be exact but to 0.1 mg. Where the sample is taken from a rolled sample, we may cut out a longitudinal strip 0.3 to 0.4 in. wide and cut this transversely into various sizes, down to mere shavings, in quantity more than sufficient for an exact weighing. This is wrapped up in 20 gm.

of silver-free lead-foil, folding it in both directions so as to form a cornet. The sample is dumped into it from the pan of the balance, and the foil folded up ready for cupelling. Cupellation is performed at a moderate temperature to form feather-litharge. Near the end, the cupel is pushed back into the muffle and finished at a higher temperature. A hot cupel should be standing ready

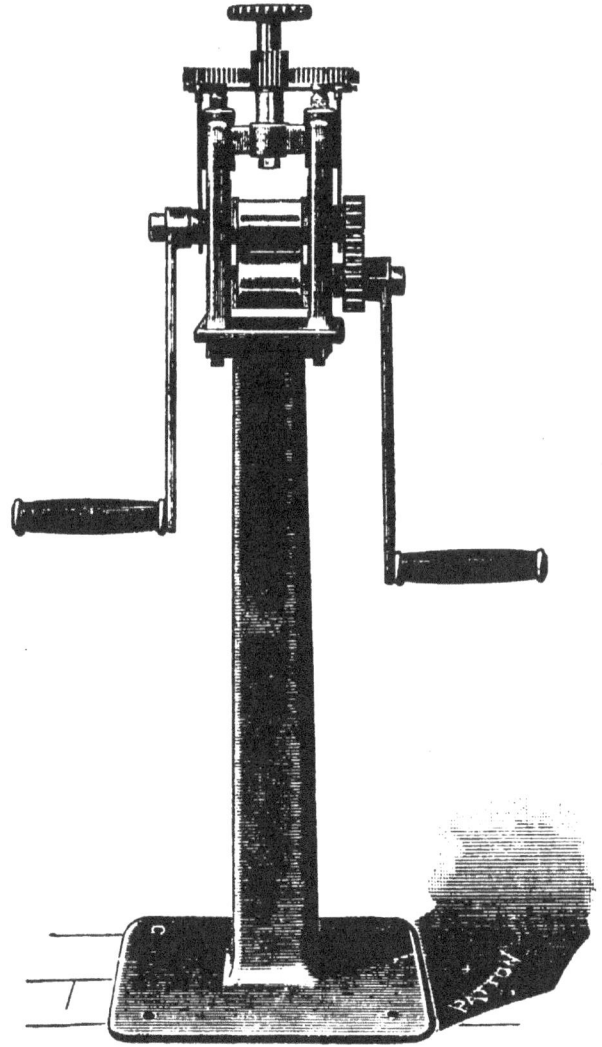

Fig. 41. Metal Rolls.

in the muffle and, at the moment of blicking, it should be put on the other, thus excluding air-currents which tend to cause so large a button to sprout or spit. This is, as its name indicates, the formation of excretions or projections upon the surface of the button due to the pressure of escaping oxygen which has just been absorbed by the molten metal, and which, upon solidification is seeking to escape. The cupel is gradually moved forward to the front of the muffle and as soon as the silver gets solid, is removed. The resultant button is weighed, and the number of milligrams multiplied by 2 will give the approximate fineness. The real fineness is greater than this by 10 to 15 fine.

PREPARATION OF THE PROOF-BUTTON. The button, which has been carefully weighed, is put in the right-hand pan of the balance, and a 500-mg. weight in the left one. Enough copper, cut from copper foil, is added with the silver button to make up the weight to 500 mg. This is transferred from the pan to a cornet of lead-foil in the manner already indicated, the quantity of foil to be taken being determined from the table herewith.

Fineness	Weight of Lead in Grams
950 to 1000	5
800 to 950	10
600 to 800	15
600 and less	20

Thus, suppose we had found by the trial-assay a button of 312 mg. equal to an approximate fineness of 624, we would see from the table that 15 gm. sheet-lead would be enough. Two other portions of the sample are taken, each of 500 mg., and wrapped in the same quantity of lead-foil as the proof. The three are cupelled abreast in the muffle as in Fig. 42, taking care to cupel them all at the same temperature and in such a way as to form feather-litharge. The cupellation loss of each of these may be considered as being the same. The fineness is, therefore, the sum of the two outer buttons plus twice the loss in cupelling. Thus suppose the results are as given in Fig. 42. Then,

$$318 + 317 + (2 \times 6) = 647,$$

which will be the actual fineness. It will be noticed that the proof is made up to imitate the sample. Strictly speaking, 5 to 10 mg. of silver should be added to the silver button to allow for the cupellation-loss, or we may take the same weight (in this case 512 + 6 = 518 mg.) of pure silver-foil. The rest of the proof

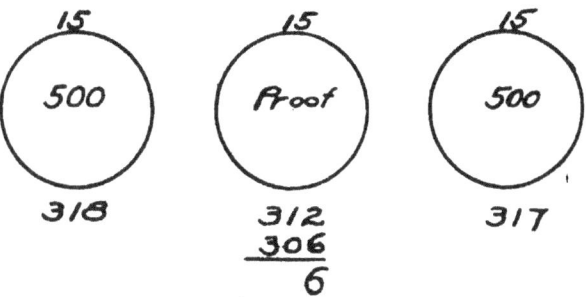

FIG. 42. CUPELLING FINE SILVER.

is assumed to be copper; but its exact amount may be determined by assay in the wet way. The duplicates should not vary from one another more than one milligram.

The gold is determined by parting the duplicates together— the weight in milligrams being the fineness, and this may be checked by parting the proof-button.

XX.—ASSAY OF BLISTER OR PIG-COPPER CONTAINING SILVER AND GOLD.

The losses of silver in attempting to assay copper matte and blister-copper by scorification are quite large, and it is accordingly best to remove the copper by a wet method before fire-treatment.

The Whitehead method of blister-copper assay is as follows: One-half of an assay-ton of the granulations is introduced into a No. 4 beaker with 50 c.c. water. To this we add 25 c.c. nitric acid of 1.42 sp. gr. and, when action has nearly ceased, another 25 c.c. is put in. The solution is allowed to stand in a warm place until the red fumes have disappeared, after which it is diluted to 300 c.c. The copper and silver have gone into solution thus:

$$3\,Cu + 8HNO_3 = 3Cu\,(NO_3)_2 + 4H_2O + 2NO$$
$$3\,Ag + 4HNO_3 = 3AgNO_3 + 2\,H_2O + NO$$

while the gold remains as an undissolved residue.

We now run into the liquid, drop by drop, a 10% solution of common salt (NaCl) in slight excess to precipitate the silver as chloride.

$$AgNO_3 + NaCl = AgCl + NaNO_3.$$

The solution is well stirred, and 10 c.c. of a 10% solution of lead acetate added and, still stirring, 2 c.c. of sulphuric acid (1 part acid to 1 part water), after which it is allowed to stand until settled. The sulphuric acid precipitates the lead thus:

$$Pb\bar{A} + H_2SO_4 = PbSO_4 + H_2\bar{A},$$

and the lead sulphate, in falling, carries down the gold mechanically. Finally the solution is filtered off, and the precipitate washed on the filter to remove copper salts. The filter paper is removed, folded over the precipitate, placed in a scorifying dish on top of 10 gm. of test-lead, and then gradually burned at the front of the muffle. A cover of 10 gm. more of test-lead, and

1 gm. of borax is added, and the whole started to scorifying. When scorified the lead buttons, weighing 6 to 8 gm., are cupelled with the formation of feather-litharge and the gold-silver button weighed and parted. The work should be done in duplicate, and the results should agree almost exactly.

The method may also be used for the determination of gold and silver in base ores, such as gray copper or arsenical pyrite, or of high-grade copper matte.

XXI.—ASSAY OF GOLD BULLION.

The sample is generally taken by two chips from opposite edges of the bar as in the silver assay. For a small bar, as little as 100 mg., and for a large one, 500 mg. may be taken.

The base metals are first removed by cupellation, wrapping the 500 mg. in 10 gm. of lead-foil. There remains behind the pure gold-silver button, which is weighed to determine the fineness of gold and silver together.

The silver is removed from the gold by inquartation, taking 500 mg. of the original sample, to which is added three times the weight of the pure silver-gold button. This is cupelled with 5 gm. of lead-foil, since it is not desired to remove all the copper. If no copper is present, 10 mg. of pure copper should be added. The resultant button is weighed, then flattened out under the hammer. It is annealed and then passed through the rolls (Fig. 43), which draws it out into a piece about four inches in length. When cold it is rolled around a lead pencil into a spiral coil, and parted in a porcelain capsule. It is boiled first for ten minutes with 1 to 1 acid (1.20 sp. gr.), then with stronger acid of 7 parts acid to 4 of water (1.30 sp. gr.) for the same period for the removal of the last traces of silver. The acids are decanted, and the residue, which is quite coherent, washed with pure hot distilled water. The gold is now dried, ignited at a low-red heat, let cool, and weighed. This weight, subtracted from the first one, gives the contained silver. The results, as before, are doubled to express the fineness.

This method is sufficiently exact for ordinary commercial assays. For the more precise methods of the government assay offices, see Furman's 'Manual of Practical Assaying,' page 246.

XXII.—ASSAY OF ORES CONTAINING METALLICS.

Ores containing native gold, silver, or copper, and by-products of metallurgical operations having metallic particles, prills, or scales, are assayed as follows:

The ore is pulverized upon the grinding plate, the metallics flattening out into scales which will not go through the screen. We accordingly have the screened material, which must be thoroughly mixed before assaying it, and the metallics remaining on the screen. Both the fines and the metallics are weighed separately.

Either one of the following methods may now be followed:

Method 1. In the total quantity, determine the percentage of metallics. Where a half-assay-ton is to be taken, figure this percentage in grams. Place that quantity on the pan of the pulp-balance, and make up to the half-ton with fines. We thus assay the whole sample in the proper relative proportions, and easily estimate the average result.

Method 2. Where the metallics are irregular in value and small in quantity, containing, say, silver, copper, and gold particles, it is better to assay the whole and determine the value in ounces per ton. Thus if the metallics weigh 7.3 gm., giving in that quantity 156 mg., we would have $\dfrac{29.166 \times 156}{7.3} = 623.3$ oz. per ton. Let us say that the fines weighed 720 gm. and assayed 115.0 oz. per ton. Then, to get the average value, we would have:

$$720.0 \times 115.0 = 82800.0$$
$$7.3 \times 623.3 = 4550.1$$

727.3   into   87350.1 = 120.1 oz. per ton.

Where the fines preponderate so greatly, we may expect this result to be but little more than the lower figure of 115 oz. per ton.

An example of an ore carrying native copper would be as follows:

The whole sample, consisting of pieces 1.25 in. diam. and smaller, weighed 142 oz. and contained 2.75 oz. of metallics which were picked out, freed from adherent gangue by hammering them, and were, by the Whitehead process (Chapter XX) found to contain 54 oz. Ag per ton.

The residue was regularly sampled, but, when finally ground to pass an 80-mesh sieve, gave 0.8 gm. metallics and 124 gm. fines.

On determination the fines gave 0.3 oz. Ag per ton and the metallics were assumed to have the same value as in the first metallics (in the 2.75 oz.).

Hence we have:

$$124.0 \times 0.3 = 37.2$$
$$0.8 \times 54.0 = 43.2$$

124.8 into 80.4 = 0.64 oz. Ag per ton.

Then, referring to the whole sample, we have:

$$139.25 \times 0.64 = 89.12$$
$$2.75 \times 54.0 = 148.50$$

142.00 into 237.62 = 1.67 oz. Ag per ton.

XXIII.—THE LEAD ASSAY.

Any lead-bearing ore may be assayed for lead, the process being a simple reduction fusion, similar to the nail-assay. Where the ore contains much pyrite it may be first roasted, as described in Chapter XIV, using 10 gm. for the roast. Ores are assayed for lead, using the following charge:

- 10 gm. ore,
- 20 gm. soda,
- 5 gm. argols,
- 4 to 15 gm. borax (used as a cover), and
- 1 to 5 ten-penny nails.

This charge, in a 10-gm. crucible, is melted down gradually so as to avoid boiling over and is finished at a high heat. As soon as it is in quiet fusion it is poured. It should not be kept long in the furnace as one may do in a silver assay, but must be poured as soon as reduction is complete.

Drops of lead, still clinging to the nails as they are withdrawn from the crucible before pouring, indicate incomplete reduction. The melt is poured into a conical mold, and, as soon as cool, the lead button at the point of the cone is removed and hammered to remove adherent slag. The button is weighed to the nearest tenth of one per cent. The average of duplicates may be taken. They should agree to within 0.5 to 1%. This fire method of lead assaying gives results a little lower than the true one, unless there are impurities as copper or antimony in the ore, which being reduced, go into the lead button. Its execution requires the best skill of the assayer, and depends on care and skill in the fire-work. The slag from such an assay pours thicker, or is stiffer, than from a silver assay. The same precautions in regard to fluxing should be taken as in a nail-assay for silver. The soda fluxes the silica and dissolves any matte which may be formed. The quantity of borax must be varied according to the amount of bases present. Lead, since it is reduced, does not enter the slag and hence is not a base that must be taken care of by borax. The nails are put in according to the needs of the sulphur, the maximum number of five being used for a pyrite or blende ore, while a single one only would be put in for an oxidized ore.

The fluxes mentioned above are often put in in mixture, the reducing mixture (Chapter VIII) having borax added to it as follows:

  20 parts monocarbonate of soda,
   5 " flour,
   5 " borax.

They are thoroughly mixed and put through a 10-mesh screen.

When thus used, the charging of the crucible is quite simple. The measured amount, say 30 gm., is put in, then 10 gm. of the ore, and both mixed with a spatula. A cover of salt, one-fourth inch deep, follows, then the needed number of nails. Additional borax may also have to be added.

Some assayers claim that in the fire-work the assay should be melted rapidly and brought to a high heat before pouring. It is well for the assayer to follow the method that gives him the best results.